UNDERKILL

SCALABLE CAPABILITIES FOR MILITARY OPERATIONS AMID POPULATIONS

DAVID C. GOMPERT | STUART E. JOHNSON
MARTIN C. LIBICKI | DAVID R. FRELINGER
JOHN GORDON IV | RAYMOND SMITH | CAMILLE A. SAWAK

Sponsored by the Office of the Secretary of Defense

 NATIONAL DEFENSE RESEARCH INSTITUTE

The research described in this report was prepared for the Office of the Secretary of Defense (OSD). The research was conducted in the RAND National Defense Research Institute, a federally funded research and development center sponsored by the OSD, the Joint Staff, the Unified Combatant Commands, the Department of the Navy, the Marine Corps, the defense agencies, and the defense Intelligence Community under Contract W74V8H-06-C-0002.

Library of Congress Cataloging-in-Publication Data

Underkill : scalable capabilities for military operations amid populations / David C. Gompert ... [et al.].
   p. cm.
   Includes bibliographical references.
   ISBN 978-0-8330-4684-0 (pbk. : alk. paper)
   1. Urban warfare. 2. Counterinsurgency. 3. United States—Armed Forces—Drill and tactics. I. Gompert, David C.

U167.5.S7U53 2009
355.4'260973—dc22

                          2009009267

Cover design by Carol Earnest
AP Photo/Emilio Morenatti

Published 2009 by the RAND Corporation
1776 Main Street, P.O. Box 2138, Santa Monica, CA 90407-2138
1200 South Hayes Street, Arlington, VA 22202-5050
4570 Fifth Avenue, Suite 600, Pittsburgh, PA 15213-2665
RAND URL: http://www.rand.org/
To order RAND documents or to obtain additional information, contact
Distribution Services: Telephone: (310) 451-7002;
Fax: (310) 451-6915; Email: order@rand.org

Preface

During the first few years of their occupation of Iraq, U.S. military forces proved to be better at killing insurgents than at defeating the insurgents by convincing the Iraqi people to turn against them. As a consequence, the insurgency grew despite its losses, the population's tolerance for the U.S. occupation shrank, and U.S. casualties mounted. At a certain point, a majority of all Iraqis believed that the use of force against U.S. troops was a legitimate form of resistance. This belief was reinforced by a number of incidents in which Iraqi noncombatants were killed or gravely hurt—cases heavily exploited by anti-U.S. propagandists. While this problem has eased in Iraq as a result of vastly improved U.S. counterinsurgency (COIN) strategy, it has called attention to the fact that U.S. forces are not well equipped to carry out operations and defend themselves amid populations except through the use of lethal force. The persistence of civilian casualties and the resulting political backlash against U.S. and North Atlantic Treaty Organization (NATO) forces in Afghanistan confirms that this deficit is a serious problem.

Against this background, a 2007 RAND Corporation report on comprehensive capabilities for COIN entitled *War by Other Means* cited, among other deficiencies, the inadequacy of U.S. nonlethal capabilities and the resultant human and political damage that comes from killing, hurting, or terrifying persons who are not enemy fighters.[1]

---

[1] See David C. Gompert, John Gordon IV, Adam Grissom, David R. Frelinger, Seth G. Jones, Martin C. Libicki, Edward O'Connell, Brooke Stearns Lawson, and Robert E. Hunter, *War by Other Means—Building Complete and Balanced Capabilities for Counterinsurgency:*

Prompted by this finding, and with the sponsorship of the Office of the Secretary of Defense, RAND embarked on a study of the requirements for and desired characteristics of nonlethal capabilities in the current and foreseeable security environments. This study was meant to take an innovative, broad, and deep look at how U.S. forces can act effectively against insurgents and terrorists without killing—and without causing lasting harm to—people among whom such enemy fighters hide. This book reports the findings of that study. It examines options for filling the void between lethal action and inaction so that U.S. troops can conduct militarily and politically successful operations amid foreign populations.

Nonlethal weapons are familiar outside and, to a lesser extent, inside the military. We all have unpleasant images of tear gas, fire hoses, batons, and rubber bullets being used against either persons considered threatening but not dangerous enough to kill or groups of people, some of whom were threatening and others of whom were not. While such crude nonlethal weapons may have their uses, this book assumes that they are inadequate for today's military missions, in part because they were not conceived for such purposes. We hypothesized that new technology (including information technology and systems whose effects can be scaled from mild to discomforting to disabling to even lethal), advanced training, and decision-making methods are required to fill this gap. Together, these technologies, training, and methods will create what we call a *continuum of force*.

Readers will find that this book resists the temptation to leap directly to intriguing technologies. An assessment of options should follow determination of need. Moreover, while technology will figure importantly in creating a continuum of force, intangible factors—cognition, operating concepts, training—should be part of the general solution. In RAND fashion, this book is interdisciplinary: The study team consisted of military-operations analysts, practitioners, policy analysts, economists, technologists, and area experts.

RAND Counterinsurgency Study—Final Report, Santa Monica, Calif.: RAND Corporation, MG-595/2-OSD, 2008.

This book should be of interest not only to persons involved in nonlethal weapons but to a much wider circle of national-security policy-makers, planners, practitioners, and scholars. We hope that this wider community will prove more aware of and committed to the development of better capabilities for military operations amid civilian populations. The prevalence of such operations across a growing set of important missions demands a higher level of attention to the issues with which this book grapples.

This research was sponsored by the Office of the Secretary of Defense and conducted within the International Security and Defense Policy Center of the RAND National Defense Research Institute, a federally funded research and development center sponsored by the Office of the Secretary of Defense, the Joint Staff, the Unified Combatant Commands, the Department of the Navy, the Marine Corps, the defense agencies, and the defense Intelligence Community.

For more information on RAND's International Security and Defense Policy Center, contact the Director, James Dobbins. He can be reached by email at James_Dobbins@rand.org; by phone at 703-413-1100, extension 5134; or by mail at the RAND Corporation, 1200 South Hayes Street, Arlington, Virginia 22202. More information about RAND is available at www.rand.org.

Contents

Figures

Tables

Summary

During COIN operations, the population is not just the field of battle but the prize of battle. Success depends on earning the cooperation of the people, whose security thus becomes one of the chief responsibilities of COIN forces. Early 21st-centrury battles have demonstrated the disadvantages faced by a force that lacks adequate options to act forcefully against insurgents without risking death or serious harm to noncombatants. In Iraq, Afghanistan, Lebanon, and, most recently, Gaza, enemy fighters have hidden in dense populations, challenging— practically daring—U.S., coalition, or Israeli forces to attack. In all three cases, superb armies with precision weapons have had to rely more or less entirely, for lack of better alternatives, on the use of deadly force against extremists who, dressed like everyone else, hide in tenements, mosques, and hospitals. The advent of global media has only compounded the problem: Enemy propagandists have a field day when COIN forces kill or injure innocent people.

The United States cannot afford to take the attitude that civilian casualties are unfortunate but unavoidable. Expressions of regret cannot repair the political damage caused by harming people whom U.S. troops are supposed to protect. When the U.S. military is entrusted with responsibility for security in another country, that country's inhabitants should be accorded the same protection from death and injury that Americans enjoy at home. A lower standard is indefensible on strategic, political, and logical grounds. In fostering effective and legitimate government in war-torn countries, the United States expects indigenous security forces to be as careful with the lives of their citizens as U.S. security services are with the lives of Americans. Because U.S. forces operating abroad must meet the same standard they prescribe for indigenous forces, the U.S. military can be no more tolerant of civilian casualties abroad than at home. Thus, for missions other than combat against identifiable enemy forces, U.S. forces should treat as paramount the safety of the people among whom they operate.

Such lofty principles will evoke some controversy. Do enemy fighters abroad have rights comparable to criminals at home? Must hostile intent be clear before U.S. troops use force? How can such a standard be reconciled with the fact that COIN may involve hostilities with persons that do not have, and arguably do not deserve, the protections accorded even the most-violent domestic criminals? Yet, these questions do not alter the fact that harming innocent persons abroad can seriously damage U.S. interests, especially when U.S. forces are responsible for the security of those very persons. This is the essence of the dilemma facing U.S. forces when they are pitted against combatants hidden among noncombatants.

Resolving this dilemma demands options that enable U.S. forces to prevail over enemy fighters without harming innocent people of similar appearance in the same location. Such options would make critical the proof of hostile intent and could neutralize dangerous individuals without harming innocent ones under U.S. protection. *Only with the right capabilities* is it possible to meet the proposed high standard of concern for innocent lives abroad without jeopardizing the missions or safety of U.S. troops in the presence of enemies with deadly intent.

The challenge of carrying out dangerous missions amid populations is not confined to COIN. Peacekeepers are often interposed between warring factions. Humanitarian-relief efforts can place U.S. soldiers in contact with desperate and unruly crowds. Intervention to halt genocide can be confounded by the mingling of predators and their prey. Quelling public disorder and rounding up looters, abroad or at home, may confront U.S. troops with the need to curb violence without using violence. The need to free hostages, isolate terrorists, and board suspicious or pirated ships with captured crews is increasing.

One is struck by the diversity of such U.S. military missions other than traditional warfare. Lumping such missions together as "irregular operations" or "stabilization operations" risks inadequate preparation for missions that can differ as much from one another as they do from regular combat. Some border on police work—e.g., enforcing order and managing crowds—which can fall to military troops when police are unavailable or inadequate, as is often the case. In other situations, U.S. forces may face enemy fighters who favor urban areas because such environments allow them to conceal themselves or endanger the people among whom they hide—or simply cause more carnage. Missions against enemy combatants usually entail different objectives, rules of engagement, and tactics than those involving the control of noncombatants. Still, the common aspect of these diverse missions— operating amid populations—points toward a general need for better options.

The growing frequency and significance of operations amid populations suggests a regular—rather than rare—need for U.S. military forces to be able to gain control of situations, perform their tasks, and protect themselves without using deadly force. Although nonlethal options have long been essential in law-enforcement missions, in which ensuring public safety with minimum violence is stock-in-trade, they have been regarded by the military as having only limited utility in only exceptional circumstances. This disregard for nonlethal weapons is evidenced by the mere $50 million spent annually on nonlethal weapons by the Joint Non-Lethal Weapons Directorate (JNLWD) of the U.S. Department of Defense. Although foreign insurgents present dangers exceeding those that police face in American cities, U.S. military forces could remedy a major shortcoming they face in COIN and other important missions if they had nonlethal capabilities that could produce a range of effects and the skills to use them. Such options would offer typical small units more flexibility, self-sufficiency, and speed; less risk of making mistakes with wide political repercussions; and better odds of accomplishing their missions.

Given the nature of the missions and responsibilities of U.S. forces, being able to disable persons without killing them is too low a standard. Even short of lethal force, violence against populations whose trust and cooperation U.S. forces need to earn—and which themselves are the key to ultimate success—may ruin the mission and set back an entire campaign. Pain, shock, or injury may turn a crowd into a mob, a mob into a confrontation, or a confrontation into a cause célèbre that can fuel insurgency. Therefore, the ability to calibrate nonlethal force from none to mild to moderate to intense can be as important as simply not causing death. The need is for a continuum of force.

In essence, this continuum must enable U.S. forces to affect the behavior of but not harm noncombatants while at the same time gaining advantage over enemy combatants who may look like and hide among those noncombatants. For example, being able to cause innocent persons and any enemy fighters who are intent on completing a hostile mission to respond in noticeably different ways would permit more-focused and more-forceful action, lethal if necessary, against the latter while minimizing harm to the former. Likewise, having the means to disorient but not injure individuals could take the initiative away from attackers without jeopardizing the well-being, good will, and future cooperation of the larger population.

To the extent possible, the continuum of force should be based on a more or less standard set of capabilities available to regular small military units involved in COIN, peacekeeping, humanitarian intervention, and other irregular operations amid populations. This need stems from the fact that the U.S. military as a rule does not rely on specialized forces for such missions but instead uses the same force types for each. The exception, special operations forces, cannot be used for every operation other than force-on-force combat. Moreover, regular units operating amid populations may not know each morning the sorts of predicaments and persons they will face that day. The need for capabilities that range from nonviolent to lethal force is common, varied, and unpredictable; the need for small units to act swiftly argues against having to call in capabilities from higher echelons.

These factors place a premium on versatile and portable capabilities that can be carried and used by small units that operate amid populations and face uncertainty. Additionally, these capabilities should be scalable—capable of producing a range of effects, from nonharmful[2] to extreme or even lethal—to enhance the ease and speed of escalation and de-

---

[2] By *nonharmful*, the authors mean harmless in intent rather than harmless in effect.

escalation as a situation unfolds as well as obviate the need for awkward or time-consuming transitions from one set of equipment to another. Furthermore, scalability implies a reduced number of different systems that may be needed, thus resulting in better portability and supportability. Continuum-of-force capabilities should also be affordable enough that most small units can be equipped with them. Finally, because missions and conditions that require a continuum of force are a present reality rather than a future possibility, technologies that are at hand or only a few years away from maturity are preferred, all else being equal, over those of speculative science.

An effective continuum of force will require that U.S. troops have

- decision-making talents that exploit information, gain time, and seize the initiative from adversaries
- performance standards and skills that allow them to escalate their use of force during a confrontation in order to gain advantage while managing risk
- readily calibrated effects that range from nonviolence to lethality.

These prerequisites can be met thanks to advances in information networking and cognition, germane experience with nonlethal force resident in the law-enforcement community, and progress in a wide assortment of potentially relevant technologies. Thus, a continuum of force is not only needed but feasible.

Having defined a general need and offered reasons to think that need can be met, we believe that specific continuum-of-force capabilities should be based on operating requirements. These requirements have been identified through examination of a diverse and representative set of realistic tactical scenarios encompassing COIN, peacekeeping, humanitarian relief, civil order, and other missions and conditions that small units might face. From 24 such scenarios, some common themes emerge:

- the prevalence of small-unit engagements and the corresponding need for junior officers and noncommissioned officers (NCOs)

to be able to decide in dangerous and urgent circumstances what measures to take
- partial, ambiguous, confusing, or deceptive information about the identity, motivations, and intentions of persons engaged
- uncertainty about the actual physical, physiological, and psychological effects of nonlethal weapon systems on individuals and groups
- the importance of seizing initiative from and exerting control over dangerous persons amid innocent ones
- the utility and difficulty of communicating with groups, especially large groups, of persons engaged by U.S. forces
- the likelihood of dire political ramifications if civilians are killed or hurt, claims of which are amplified by global media and distorted by enemy propagandists
- the need for mild, even nonviolent, initial effects in order to gain time, information, initiative, and control, including the differentiation or separation of combatants from noncombatants
- the importance of rapidly scalable and portable capabilities.

These findings suggest that the problem of acting forcefully against adversaries amid populations is as much one of gaining and using knowledge as causing desired effects. Therefore, an integrated solution—part information, part judgment, and part physical capability—is needed, and has been missing so far in the search for nonlethal options.

In regard to physical capability, we assessed numerous options using operating requirements derived from the study's scenarios, along with four key general criteria: versatility, portability, scalability, and feasibility. Some options appear to be efficacious under some but not all conditions. For example, a small unit patrolling neighborhoods without knowing whether, when, and what threats could appear cannot routinely include in its patrols a heavy truck with a microwave transmitter. Anti-electronics capabilities are of value only when enemy fighters are in vehicles or otherwise relying on electronics. Flash-bang munitions have limited range and may be frightening to innocent persons who happen to be present. Electric-shock tasers are useful only at short range against small numbers of individuals, are not scalable, and can cause pain. Tear-gas may alienate otherwise sympathetic persons. Rubber bullets produce pain, if not injury, and are not scalable. Recognizing that these options may be useful only in specific circumstances, we aim to identify options with such wide utility across possible missions and conditions that ordinary small units should and could routinely be equipped with and trained to use them.

Although no single technology satisfies this general need, the options of greatest promise involve sound and light. Both can be effective in hailing, inhibiting, disorienting, disrupting, or degrading the key sensory faculties of dangerous persons up to hundreds of yards away without necessarily causing lasting harm to them or to innocent persons. Their effects can range from mild to severe, affording users the opportunity to observe the effects of their weapons and make adjustments. As an effect's intensity increases, enemy combatants and noncombatants may react differently, isolating the former and scattering the latter, thus reducing the number of potential targets against which to direct even harsher measures. Sound and light can be used against large groups, small groups, or individuals. Considering a wide range of lasers—from low-energy to high-energy to femto-second pulses—means that the desired effects can be even more pronounced.

Directed sound and light, including light from lasers, appeared useful in many of the study's scenarios and thus are versatile. They are sufficiently portable (on foot or in small vehicles) that platoons or squads can carry them on most missions. They do not involve physical projectiles, violent contact, or inhalation, any of which could prove counterproductive if used against people whose cooperation is needed for the mission to succeed.

As part of an integrated approach, directed-energy capabilities can be complemented by an innovative communications technique involving the use of cell phones. If friendly authorities have privileged access to cell-phone switches, a unit commander could request that all cell phones in a given neighborhood or congregated group of people be called to transmit simple text or audio messages that alert, warn, reassure, or instruct. Even if only a fraction of those present receive the message, the rest would be promptly told its contents. The utility of adding this feature to a directed-energy suite of capabilities was apparent in many of the study's scenarios.

In addition to a cell-phone warning system, the continuum of force could exploit video technology. Vehicle-, weapon-, or fixed-mounted cameras and video recordings could aid in learning lessons, adapting systems and tactics, refuting unfounded rumors and propaganda, and collecting intelligence about, for instance, dangerous persons. In addition, live video could help forces manage escalation or de-escalation.

An assessment of technologies cannot be based on technical grounds alone. Military operations amid populations are fraught with political risks, which enemies and unfriendly media organizations are poised to exploit. The potential for adverse reactions among people affected or the wider population is a function of both the severity and the strangeness of the effects of a given capability. The unfamiliar may give rise to panic, rumor, superstition, and disinformation. However painful their effects, rubber bullets at least will not be blamed for subsequent tumors, impotence, infertility, or mental disorder. Even intense sound and light are less likely to cause adverse psychological and political reactions than are chemicals, shocks, or "rays." At the same time, the use of lasers might be misunderstood by those illuminated or misconstrued by propagandists.

In the same vein, cell-phone messaging to alert and inform citizens about the use of nonlethal force may raise psychological and political questions. Although citizens may appreciate being warned, instructed, or reassured, they may at the same time react adversely to the perception that U.S. forces or their own government is able to send them messages at will and, by implication, access their phones (and conversations). One way to win public acceptance for the cell-phone messaging concept is to give each person a choice of whether to subscribe to public warnings upon acquiring a cell phone or service contract. Although some would decline the option, those who did subscribe would most likely not be suspicious upon receiving an alert message—in fact, they would be reassured. In time, more people would likely sign up for this service. Likewise, people suspicious of increasing levels of video-camera surveillance would have to be educated about and convinced of the security benefits.

In any case, it is imperative to communicate early, persistently, and accurately the rationale behind and facts of all aspects of a continuum of force to people who may be affected. The unifying theme of such communication must be that U.S. forces accept their duty to safeguard the people of countries where they operate and, accordingly, are depriving killers of the benefit of hiding among and harming those people. Without such communication, even careful use of nonlethal force can go awry.

The suite of capabilities described in this book is for the most part technologically feasible. Aspects that require development include the following:

- very-high-intensity[3] sound that is precise, scalable, effective at long ranges (i.e., hundreds of yards), and can cause discomfort, disorientation, or incapacitation
- femto-second lasers
- software that permits selective and instantaneous cell-phone messaging to users in a particular area
- deployable links for real-time video
- improved portability of all elements of the suite, with a view toward fielding some or all capabilities with dismounted troops, thus improving versatility.

In addition, capabilities should be engineered as an integrated system suite with, for example, common power sources, displays, controls, and physical packaging.

Of course, the requirement for an integrated suite adds to complexity and raises concerns about the feasibility of the whole. A more serious potential problem than the feasibility of individual pieces themselves is whether the integration of the components, including important information and communications features, is feasible in the near-to-medium term, especially when taking into account the need for sophisticated operators and nuanced doctrine. Without underestimating the associated challenges, we regard such integration as well

---

[3] That is, powerful and focused.

within the capability of the U.S. military and its system providers. Moreover, the suite proposed here does not depend on, and should not await, every component.

In its fullest form, then, the suite of continuum-of-force capabilities envisioned here could consist of directed sound, directed light, lasers, cell-phone communication, and video observation. As a whole, this suite is remarkable in its nonkinetic character: For the most part, it affects the senses and perceptions rather than the physical condition of persons engaged. This does not mean that kinetic nonlethal or lethal capabilities have no place in the range of options available to U.S. forces operating amid populations. Yet, the idea of alternatives to physical violence leads to a host of emerging but largely proven technologies which, used creatively and together, offer U.S. forces ways to control situations and gain advantages over enemies without harming persons who ought not to be harmed.

As important as developing and integrating technology is ensuring that this nonlethal capability can be incorporated into and used effectively by ordinary small U.S. military units. A continuum of force must include abilities to sense and grasp a fluid situation, judge how to respond when the nature and intentions of the persons engaged are unclear, understand and anticipate behavior, communicate, escalate and de-escalate wisely, and be sensitive to the wider repercussions of actions. To use well the technologies suggested here, it is necessary to instill methods of adaptive decision-making under pressure.

To incorporate continuum-of-force capabilities into an ordinary small unit, it would be better to rely on a well-trained, experienced, specialized *team-within-unit* than to prepare, involve, and have to control every member of the unit. The former approach would allow the unit commander to concentrate on the essential tasks of sensing, reasoning, and adapting during the operation. Placing high-powered sound, light, and lasers in one vehicle fits with the team-within-unit approach. Finally, a team-within-unit would allow most members of the unit to be concerned only with the use of deadly force, thus lowering the risk that nonlethal options might impede the use of lethal ones.

Assuming such an approach is adopted, the military will need to invest in the requisite human resources and abilities, especially

- training and educating junior officers and NCOs in continuous sense-making and decision-making when faced with uncertainty, urgency, and risk—physical and political—amid populations
- selecting NCOs with the temperament and aptitude for technical and behavioral aspects of the continuum.

A related matter is the nature and content of instructions to be issued to these teams-within-units and their commanders. A notable advantage of relying on a few well-trained and seasoned NCOs is that they will not require detailed or rigid instructions. Given the uncertainty and fluidity of situations requiring a continuum of force, clear but flexible guidelines akin to those on which police departments rely are preferable to elaborate field manuals and checklists.

Creating a continuum of force will require a multifaceted effort that is best conducted by one of the U.S. military services acting as executive agent. There is no compelling reason why the Marine Corps should relinquish its current role as executive agent unless, upon considering future missions, it concludes that the continuum is not crucial for its small units. In that event, tempting as it is to look to Special Operations Command to introduce innovative capabilities, it must be remembered that the situations in which a continuum of force may be needed are so common that regular ground-force units must be prepared to use the continuum. This argues for making the Army the executive agent if the Marine Corps declines the role. Alternatively, given that several services could make use of continuum-of-force capabilities, a case can be made for placing the responsibility with Joint Forces Command.

In any case, the scope of JNLWD's work should be expanded beyond nonlethal technology to include sensing, cognition, and communications. Considering this requirement and the need for research and development of the suite of capabilities suggested here, we recommend an additional $250 million in funding for JNLWD for 2009–2013, roughly doubling its current budget. More funding than that will be needed, of course, as new capabilities are acquired.

As the U.S. military fashions a continuum of force, we urge it to pursue international collaboration, not only with close U.S. allies (e.g., NATO) but also with the United Nations' peacekeeping department and a wide circle of like-minded countries with similar needs. There are few if any risks associated with such collaboration, and it is in the U.S. interest to foster widely the fielding of capabilities that can be effective against enemy fighters without harming civilians.

In sum, a continuum of force for regular U.S. troops operating amid populations is needed and possible. Scalable and portable technologies—e.g., directed sound and light—are in train or within reach. But those technologies do not provide a complete solution: The ability to prevail against dangerous enemies without harming innocent people and jeopardizing larger campaign goals depends crucially on the skill, sensitivity, and preparation of U.S. soldiers. In turn, creating and mainstreaming this ability will require vision, initiative, commitment, and persistence on the part of those soldiers' civilian and military leaders.

Acknowledgments

The authors would like to thank the numerous individuals who supported and contributed to this research. The study on which this book is based was made possible by the valuable assistance of Benjamin Riley, Director, Rapid Reaction Technology Office; Mark Gunzinger, Deputy Assistant Secretary of Defense, Forces Transformation and Resources; and Terry Pudas, former Deputy Assistant Secretary of Defense for Forces Transformation and Resources.

We appreciate the important insights provided by the following individuals and organizations: Joint Non-Lethal Weapons Directorate, Quantico, Va., particularly Scott Griffith, Carlton Land, David Law, Richard Scott, and Kevin Swanson; Wes Burgie, American Management Systems, Inc.; Andrew Hughes, Department of Peacekeeping Operations, United Nations; Colonel Ian Sinclair, Department of Peacekeeping Operations, United Nations; and Col. Michael Frazier (U.S. Marine Corps) of the Rapid Reaction Technology Office.

We also thank Daniel Gouré of the Lexington Institute and Steve Berner of RAND for their incisive reviews, which made this book better. We recognize our RAND colleagues, Kayla Williams, Omar Al-Shahery, and Rebecca Bouchebel, who aided the research.

Abbreviations

ADS	Active Denial System
AFRICOM	Africa Command
COIN	counterinsurgency
CONOP	concept of operation
DoD	U.S. Department of Defense
HMMWV	high-mobility multipurpose wheeled vehicle
HVT	high-value target
ICON	integrated counterinsurgency operating network
IED	improvised explosive device
JNLWD	Joint Non-Lethal Weapons Directorate
LNG	liquefied natural gas
MP	military police
NATO	North Atlantic Treaty Organization
NCO	noncommissioned officer
NYPD	New York City Police Department
R2P	responsibility to protect
RoE	rules of engagement

SOCOM	Special Operations Command
SOFA	status-of-forces agreement
TCP	traffic control point
UN	United Nations
WTO	World Trade Organization

CHAPTER ONE

Introduction

Framing the Challenge

During counterinsurgency (COIN) operations, the population is not just the field of battle but the prize of battle. Success depends on earning the cooperation of the people, whose security thus becomes one of the chief responsibilities of COIN forces. Early 21st-century battles have demonstrated the disadvantages faced by a force that lacks adequate options to act forcefully against insurgents without risking death or serious harm to noncombatants. In Iraq, Afghanistan, Lebanon, and, most recently, Gaza, enemy fighters have hidden in dense populations, challenging—practically daring—U.S., coalition, or Israeli forces to attack. In all three cases, superb armies with precision weapons have had to rely more or less entirely, for lack of better alternatives, on the use of deadly force against extremists who, dressed like everyone else, hide in tenements, mosques, and hospitals.

From the Balkans to Iraq to Afghanistan, U.S. troops have been increasingly, almost continuously, engaged in military operations amid populations among which enemy fighters conceal themselves, recruit, plot, prepare, and strike. Such populations have been friendly, ambivalent, or antagonistic toward U.S. troops, and often some of each. Even to trained eyes and advanced sensors, distinguishing enemy fighters from unfriendly protesters or innocent onlookers may be impossible. Blending into a population—inviting attack—is a favorite enemy tactic and a successful one insofar as U.S. forces cannot take forceful action without endangering noncombatants.[1]

The core definition of *insurgency*—an armed challenge to the government for the people's allegiance—applies throughout the troubled Muslim world, where shaky regimes battle extremist groups with substantial followings. It is not only the difficulty of using force against enemies lurking in populated areas but also the risk of antagonizing contested populations that increasingly preoccupy U.S. troops and leaders. With the advent of global media, anti-American propagandists have a field day when U.S. soldiers kill or injure noncombatants.

The damage from adverse publicity—experienced time and again in Iraq and Afghanistan—is a reminder that COIN, like all conflicts, is fundamentally and ultimately political. As Carl von Clausewitz observed, "the political object—the original motive for war—will thus determine both the military objective to be reached and *the amount of effort it requires*."[2] In COIN, the epicenter of politics is the contested population. Failure to treat that population with care, even reverence—as good police treat even bad neighborhoods—will affect and possibly determine a campaign's course and outcome. This is the light in which the utility of nonlethal weapons appears not only tactical and operational but strategic.

While such conditions are a hallmark of COIN, they are not confined to COIN. Peacekeepers are often interposed between warring factions. The delivery of humanitarian-relief supplies may place U.S. soldiers into contact with demanding and unruly crowds. Intervention to halt genocide can be complicated by the mingling of predators—often teenagers—with their prey. Quelling public disorder and rounding up looters, usually abroad but possibly at home, may confront U.S. troops

---

[1] At the time of this writing, concern is mounting about the frequency and consequences of noncombatant casualties caused by U.S. air strikes (especially in Afghanistan). While this is a problem in need of attention, such as through better intelligence, targeting, precision, and discretion, the main focus of this study was on ground (and to some extent maritime) missions.

[2] Carl von Clausewitz, quoted in Timothy J. Lamb, "Emerging Nonlethal Weapons Technology and Strategic Policy Implications for 21st Century Warfare," *Military Police*, PB 19-03-1, April 2003, p. 8., emphasis added.

with the challenge of curbing violence without resorting to violence. The need to board suspicious or pirated ships with captured crews is growing. Rescuing hostages and stopping suicide terrorists before they detonate their vehicles or explosive belts in markets or religious gatherings present increasingly acute problems.

Such missions are no longer of secondary importance to a U.S. military designed to wage regular war against regular armies. Taken together, they are of equivalent importance—and far more frequent.[3] They raise a common problem: how to exert control without causing loss of life or limb. Long a challenge faced by police, this issue is now confronting military troops, who need better capabilities—tools as well as skills—to gain advantage, carry out their missions, and protect themselves without harming civilians. Although the U.S. military has invested in and used so-called nonlethal weapons, avoiding the death of noncombatants is not an adequate standard. *Any* violence against people whom U.S. forces are expected to safeguard and win over can undermine a mission or an entire campaign.

The U.S. military is revising its operating concepts, training, and equipment to succeed at COIN. Yet, recent RAND research finds that U.S. forces in COIN operations need the ability to gain control of a given situation or group of people without killing or injuring them.[4] If the U.S. military plans to give this deficiency the attention it merits, it must rethink two basic presumptions: first, that U.S. troops must use deadly force to carry out military missions, and second, that civilian casualties are bound to happen. The first presumption is untrue in many missions other than force-on-force combat. The second implies

---

[3] Department of Defense, Directive 3000.05, *Military Support for Stability, Security, Transition, and Reconstruction Operations*, November 28, 2005. This document is a Department of Defense (DoD) policy statement that raises stability operations and irregular warfare to the level of regular warfare in importance.

[4] David C. Gompert, John Gordon IV, Adam Grissom, David R. Frelinger, Seth G. Jones, Martin C. Libicki, Edward O'Connell, Brooke Stearns Lawson, and Robert E. Hunter, *War by Other Means—Building Complete and Balanced Capabilities for Counterinsurgency: RAND Counterinsurgency Study—Final Report*, Santa Monica, Calif.: RAND Corporation, MG-595/2-OSD, 2008.

an acceptance of the inevitability of civilian casualties that carries grave operational, human, and political risks.

When the U.S. military is entrusted with responsibility for security in another country, the inhabitants should be accorded the same protection from death and injury that Americans enjoy at home. A lower standard is indefensible on strategic, political, and logical grounds. In fostering effective and legitimate governments in such countries as Iraq and Afghanistan, the United States wants indigenous security forces to be as careful with the lives of their citizens as U.S. security services are with the lives of Americans. Because U.S. forces operating abroad must meet the same standard prescribed for indigenous forces, U.S. forces in such missions should be no more tolerant of death and injury among innocent civilians abroad than at home.

This insight will surely evoke some controversy, as it should. One high-ranking United Nations (UN) peacekeeping officer states that it is the right perspective, and one that the UN tries to impart to its personnel and operations.[5] Others ask how the precautionary standard proposed here would apply to insurgents with deadly intent hidden amid populations.[6] Do enemy fighters abroad have rights comparable to criminals at home? Must hostile intent be clear before violent or lethal force is used by U.S. troops, as is the case with police or military forces (e.g., the National Guard) operating within the United States? How can such a standard be reconciled with the fact that COIN may involve combat with persons that do not have, and arguably do not deserve, the sort of protection accorded even the worst domestic criminals?

Still, these questions do not alter the fact that harming innocent people abroad can seriously damage U.S. interests, especially when U.S. forces are responsible for the security of those people. This is the essence of the dilemma facing U.S. forces pitted against enemy combatants hidden among noncombatants. Resolving this dilemma requires a range of capabilities that enable U.S. forces to prevail over their enemies without harming innocent people of similar appearance in the vicin-

---

[5] Author interview with a senior official, Office of Military Affairs, UN Department of Peacekeeping, New York, November 20, 2008.

[6] Daniel Gouré, one of the reviewers of a draft of this book, November 2008.

ity. Such capabilities make proof of hostile intent less critical because their effects result in less than death or serious harm to individuals who may turn out not to be hostile.[7] Moreover, having options that can neutralize dangerous individuals without harming the innocent would improve the ability of U.S. forces to safeguard noncombatants. *With the right capabilities*—the object of this study—it is possible to meet the proposed high standard of concern for innocent lives abroad without compromising the missions or safety of U.S. troops, even in the presence of enemy fighters with deadly intent.

Thus, in COIN and similar circumstances, U.S. forces must be able to use deadly or disabling force, just as they must be able to use nonharmful force.[8] *Force* should be thought of as a continuum from nonviolent to lethal. Minimizing violence, a norm of law enforcement, must be at least an option in military operations. At the same time, the availability and use of capabilities that do not kill or injure must not compromise the ability or preparedness of U.S. forces to take deadly action when they must, which has been a long-standing concern of military leaders vis-à-vis nonlethal weapons.

Although traditional nonlethal weapons—e.g., rubber bullets and tear gas—are useful under certain circumstances, they neither begin to fill the need for a continuum of force nor are broadly useful. Moreover, individuals who have been shot with rubber bullets or have gagged on tear gas may not upon recovering be inclined to cooperate with the troops who have used these weapons on them. Success in COIN requires not merely a passive but an actively cooperative population.

There is thus a need for innovative solutions that combine advanced technology with meticulous training and refined decision-making so that the ordinary small U.S. military unit can access a full range of force options that afford it decisive advantages over enemy fighters while sparing the lives, well-being, and cooperative disposition of the popula-

---

[7] This is the same logic that attracts police in the United States to non-lethal options that can be used to protect their own forces and ordinary citizens without waiting for evidence of hostile intent or violating the rights of potentially dangerous persons who have yet to commit a crime.

[8] By *nonharmful*, the authors mean harmless in intent rather than harmless in effect.

tions within which those enemies operate. Military forces, in contrast to police, have regarded such a capability as an exception needed under unusual circumstances rather than a general requirement. The need has been increasingly recognized in recent years, but the lack of emphasis on and resources for nonlethal capabilities (especially nonviolent ones) reveals a failure to appreciate fundamentally that a continuum of force is essential for depriving enemies of the advantages of waging, in General Rupert Smith's words, "war amongst the people."[9]

In sum, with COIN, peacekeeping, humanitarian intervention, and other irregular missions increasingly common and likely to remain so, this book questions the presumption that U.S. military forces should rely on deadly violence except in rare circumstances. More than that, it explores what capabilities are needed to form a continuum of force and what it will take to develop, field, and use such capabilities. To that end, the book

- examines how U.S. policy and strategy shape the need for a continuum of force[10]
- lays out generally the elements that could and should constitute such a capability
- analyzes a diverse set of operating scenarios to determine requirements
- identifies and assesses interesting options for meeting these requirements
- defines the capabilities that appear most promising as a general solution for the ordinary small unit
- analytically tests these capabilities in the same scenarios used to generate requirements for the purpose of identifying their impact on results

---

[9] Rupert Smith, *The Utility of Force: The Art of War in the Modern World*, New York: Knopf, 2007.

[10] Although focused on U.S. requirements, this book recognizes that other leading nations and international security organizations (e.g., the UN and NATO) have similar requirements. Moreover, with its role in peacekeeping and police operations, the UN has some experience that bears on the question.

- suggests a notional concept of operations and implications for small-unit organization
- addresses integration, feasibility, and implementation, and makes recommendations.

This book on the study's results is structured on these lines.

Current DoD Nonlethal Weapons Programs

Although a premise of this study is that U.S. military forces lack adequate options for forceful action amid populations, current DoD efforts in this realm need to be noted. Nonlethal weapons are defined by the DoD as "weapons . . . designed and primarily employed so as to incapacitate personnel or materiel while minimizing fatalities, permanent injury to personnel, and undesired damage to property and the environment."[11] They are intended to incapacitate their targets, have reversible effects, and limit collateral damage and undesired effects. Desired nonlethal weapon effects are to discourage, delay, or prevent hostile actions; limit escalation; permit military action when use of lethal force is not the preferred option; protect U.S. forces; and temporarily disable enemy equipment, facilities, and personnel.

Responsibility for nonlethal weapons within DoD lies with the Joint Non-Lethal Weapons Directorate (JNLWD). Established in 1997, the directorate is the focal point of nonlethal weapons–related research and development. Atop JNLWD, the U.S. Marine Corps is the DoD Executive Agent for the nonlethal weapons program. To date, the JNLWD has developed and fielded nonlethal weapons for the Army, the Marine Corps, the Navy, the Air Force, and the National Guard in two core areas: counterpersonnel and countermateriel (i.e., against vehicles, vessels, aircraft, buildings, etc.). More than 40 types of nonlethal weapons are currently fielded, including high-intensity

---

[11] U.S. Army, U.S. Marine Corps, U.S. Navy, and U.S. Air Force, FM 3-22.40/MCWP 3-15.8/NTTP 3-07.3.2/AFTTP(I) 3-2.45, *NLW Multi-Service Tactics, Techniques, and Procedures for the Tactical Employment of Nonlethal Weapons*, Washington, D.C., October 2007, p. 2.

light, microwave weapons, flash-bang munitions, lasers ("dazzlers"), anti-electronic devices, tasers, sound arrays, pepper spray, and blunt-impact munitions.

Some of the newer nonlethal weapons under development include

- *acoustic hailing devices*, which use advanced directed-sound beams to provide a warning capability at a greater range than existing nonlethal systems
- *Improved Flash-Bang Grenades*, which increase temporary incapacitation and improve the effectiveness and safety of currently fielded nonlethal flash-bang munitions
- Vehicle Lightweight Arresting Devices, which are portable, pre-emplaced nets equipped with a barbed spike system designed to stop vehicles traveling at high rates of speed
- Joint Non-Lethal Warning Munitions, which are small-arms cartridges that can project clear, unambiguous warning signals at distances of 100 meters, 200 meters, and 300 meters
- *Airburst Non-Lethal Munitions*, which enable precision airburst delivery of nonlethal weapon munitions
- *Mission Payload Module—Non-Lethal Weapons Systems*, which are mounted on high-mobility multipurpose wheeled vehicles (HMMWVs) to field an array of nonlethal munitions
- *optical distractors*, which are visible-laser devices that use highly directional optical energy at long ranges and have a reversible optical effect
- stopping devices, which are capabilities such as high-power microwaves, high-energy lasers, and direct electrical injections that stop vehicles and vessels at greater ranges without the need for pre-emplacement.

These efforts reveal at least some DoD recognition of the problem of using force without harming and alienating the populations amid which U.S. units operate. Moreover, this study found that JNLWD is capable, innovative, and dedicated to finding solutions. However, JNLWD's budget is a mere $50 million per year—a minute fraction of the DoD's $80 billion research and development budget and negligible when weighed against the potential benefits of improving U.S. capabilities for a host of important missions other than force-on-force warfare. Moreover, this study did not uncover elsewhere in DoD great interest in this problem, let alone a strong commitment to solving it. Thus, while taking into account JNLWD's work—and having benefited greatly from that office's cooperation and ideas—we believe it is fair to say that the U.S. defense establishment has gotten no farther than the foothills of surmounting this formidable challenge.

Concern about lack of DoD support for the improvement and employment of nonlethal weapons has appeared repeatedly and prominently in independent assessments in recent years. For instance, in 2004, the Council on Foreign Relations "found little evidence that the value and transformational applications of nonlethal weapons . . . are appreciated by the senior leadership of DoD" and concluded that although the utility of nonlethal weapons "has been demonstrated when they have been used, non-lethal weapons are not yet widely integrated into [U.S.] armed forces."[12] The National Research Council of the National Academies determined in 2003 that the "current scope of the program offers only a low probability of moving even the best ideas to the field in the future."[13] In an April 2003 article, Paul K. Shupe argued that (1) the United States does not have adequate "concepts, training, and application of non-lethal weapons"; (2) "there is no satisfactory national guidance or strategy that clearly defines . . . the importance of non-lethal weapons"; and (3) "DoD does not have appropriate joint organizational hierarchy with adequate resources to develop non-lethal weapons capabilities."[14] And in 2006, Jeffrey L. Underhill wrote that "DoD's . . . 'kinetic culture' inhibits the development . . . of

---

[12] Council on Foreign Relations, *Non-Lethal Weapons and Capabilities: Report of an Independent Task Force*, New York: Council on Foreign Relations, Inc., February 2004, p. 1.

[13] National Research Council of the National Academies, Committee for an Assessment of Non-Lethal Weapons Science and Technology, *An Assessment of Non-Lethal Weapons Science and Technology*, Washington D.C.: The National Academies Press, 2003, p. 7.

[14] Paul K. Shupe, "Nonlethal Force and Rules of Engagement," *Military Police*, PB 19-03-1, April 2003, p. 43.

non-lethal weapons."[15] Even DoD's own Joint Requirements Oversight Council is on record as having said that

> the U.S. military lacks the ability to engage targets that are located or positioned such that the applications of lethal, destructive fires are prohibitive or would be counter-productive to the U.S. objectives and goals. Operational and strategic applications of non-lethal weapons *do not exist*. . . . [T]he U.S. needs a non-lethal capability that can help defuse volatile situations, overcome misinformation campaigns, and break the cycle of violence that often prolongs or escalates conflict.[16]

What is especially notable about this statement is that it was made *before* the U.S. engagements in Afghanistan and Iraq and, yet, is no less true today.

In sum, there is a sizable and widening discrepancy between the importance of nonlethal weapons in the 21st century and the recognition of this fact as indicated by the attention and resources being devoted to it by the U.S. defense establishment. Ironically, commanders returning from COIN operations often report on this discrepancy, yet, nonlethal weapons are still not being integrated into the military's mainstream. This study not only confirms this finding but also offers a conceptual focus and practical ideas to give U.S. forces options they need to achieve both the military and larger political objectives of their missions.

---

[15] Jeffrey L. Underhill, *Are the Department of Defense Non-Lethal Weapon Capabilities Adequate for the 21st Century?* U.S. Army War College Strategy Research Project, Carlisle Barracks, Pa.: U.S. Army War College, 2006, p. 2.

[16] Council on Foreign Relations, *Non-Lethal Weapons and Capabilities*, 2004, p. 3, emphasis added.

CHAPTER TWO

The Policy Setting

The Void Between Lethality and Inaction

American generals like to say that the purpose of U.S. forces is to fight and win the nation's wars. But they and the rest of us know that nowadays it is not that simple. In the current security environment, conflict with the armed forces of other states is but one of several important missions.[1] Since the Gulf War of 1990–1991, U.S. forces have engaged in protracted COIN campaigns, postconflict stabilization and reconstruction operations, clandestine searches for terrorists, humanitarian intervention, multilateral peacekeeping, local security-force capacity-building, domestic and foreign disaster relief, and a variety of other missions. These missions differ markedly from force-on-force combat, yet, taken together, can no longer be regarded as exceptions. In such circumstances, U.S. forces, having been designed and equipped for combat, have one predominant way to exert their will: using, or threatening to use, deadly force.

To say that U.S. forces depend predominantly on deadly force does not mean that they depend on indiscriminate force—far from it. Thanks to joint collaboration and information networking, U.S. forces can now operate in integrated ways, giving them multiple lethal options (e.g., air strikes, ground fire, missiles, and special operations),

---

[1] In 2007, stabilization operations were elevated to equal importance with combat in DoD guidance.

often in the same place and at the same time. Thanks to new technology, U.S. forces can exploit precise sensing, calibrate the scale of attack, strike with high accuracy, and minimize collateral destruction and death. And thanks to better preparation for missions other than regular combat, U.S. troops are trained to use force judiciously.

The problem is not that U.S. forces lack care in using force but that they lack options short of lethal force. As a consequence of this deficiency, errors too often result in killed and seriously wounded noncombatants. Usually, the alternative to using deadly force is to use no force at all, which can endanger U.S. objectives and troops. Even when U.S. forces rely on their persuasive powers, psychological operations, and other nonkinetic means to gain control of people and situations, the threat of deadly force is implied. The premise of this study is that the use and threat of deadly force are essential but insufficient to assure success in many important missions U.S. forces will be asked to perform, including COIN.

Some missions do not require lethal force to succeed. If the aim is to maintain public safety, defuse civil disorder, distribute relief supplies, or separate opposing factions, it may be that people need to be warned, moved along, stopped, held, sorted, searched, or questioned but not necessarily subjected to violence. When enemy fighters are believed to be present but cannot be distinguished from ordinary people, the use of deadly force can be unproductive or counterproductive. Yet, failing to act purposefully because of such sensitivities and risks could imperil a mission's goals, the lives of U.S. troops, and the careers of U.S. commanders.

In sum, for many of the demands on U.S. forces today, there is a gaping capability void between deadly force and no force. The consequence of this void is not just that U.S. forces may on unfortunate occasions cause civilian casualties—which is bad enough—but that they may not be able to operate effectively *in general* against enemies amid populations. According to a former deputy commandant of the Marine Corps, nonlethal capabilities can

> fill a vulnerabilities gap, and, in so doing, allow American forces
> to effectively address a wider variety of situations and better con-

trol the escalation of violence in many situations. Non-lethal capabilities make our forces more, not less, formidable.[2]

Missions *Sans* Martyrs

Anyone doubting the significance of this void between "kill" and "do nothing" should consider the difficulties U.S. forces have experienced in securing urban areas of Iraq and Afghanistan. Even when no mistakes are made, deadly action by U.S. forces is exploited to support the jihadist claim that Muslims are under attack by modern-day "crusaders." When mistakes are made—weddings strafed, women and children injured, homes destroyed—the images are transmitted worldwide via the Internet, satellite TV, and cell-phone photos, stoking anger and calls for vengefulness against the United States, its policies, and its military presence in Muslim lands. Even without jihadist propaganda, media scrutiny ensures that news of violence against civilians by U.S. forces will spread and trigger demands for—if not acts of—retribution. Yet, inaction can have equally deleterious effects: U.S. authority weakened, local populations shown that insurgents have the upper hand, the authority of local government discredited, enemies emboldened.[3]

The void between lethal force and no force at all can be especially consequential when enemies are known to be present but look like ordinary people, or when a group of people could turn from anger to violence. Such problems are most pronounced in urban areas due to high population density and the difficulties U.S. forces experience in gaining accurate awareness, communicating, and maneuvering. Because operating in urban areas is a common and demanding case, it is a focus of this study.

---

[2] E. R. Bedard, "Nonlethal Capabilities: Realizing the Opportunities," *Defense Horizons*, No. 9, National Defense University, Center for Technology and National Security Policy, Washington, D.C., March 2002.

[3] In Iraq, the dilemmas involving the use of deadly force have been if anything harder to solve with Shiite militia which, more so than Sunni extremists, often fulfill local security functions, cluster at religious sites, and are appreciated by the people.

One of the cardinal mistakes in COIN is to believe its chief aim is to slay insurgents. Rather, the chief aim of COIN is to convince people to support their government, assuming it is worthy of their support. Depending on circumstances, slaying insurgents may or may not serve this goal. To the extent that the insurgents are feared or hated by the population, eliminating them helps. But to the extent a sizable portion of the population sympathizes with the insurgents' cause and sees them as champions, slain insurgents become glorified martyrs, swelling the stream of recruits to take their place. Such effects can be much stronger when military operations are being conducted by foreign ("infidel") forces.

Just as insurgency has changed with the advent of global communication and collective identities (e.g., the global "Muslim Nation," or *Ummah*), so must COIN change. The pattern for 21st-century insurgency has been set by the spread of organized radical Islamist violence against the West and its proxies in the Muslim world. Al Qaeda, among others, feeds and feeds on local insurgency, making each insurgency more dangerous and also self-replicating. Such insurgencies are not conducive to political accommodation and cannot be defeated by conventional military means alone. They rely heavily on a particularly destructive, fear-inducing, and indefensible weapon: the suicide bomber. They transmit both their own feats and U.S. blunders to the *Ummah*, increasing its disaffection toward existing regimes and distrust of the United States. They have depicted the U.S. "global war on terror" as the latest wave of Christian assaults on Islam going back a thousand years. Their source of energy is the ability to convert alienated Muslims into jihadists by deepening and capitalizing on a mindset of victimhood and self-defense.[4] Muslim casualties, whether innocent or not, are the currency of the call to holy war.

While it may or may not be productive to kill insurgents, it is decidedly counterproductive to kill noninsurgents. Whatever the population's attitude about the insurgents, the chances of the government

---

[4]   John Mackinlay and Alison Al-Baddawy, *Rethinking Counterinsurgency: RAND Counterinsurgency Study—Volume 5*, Santa Monica, Calif.: RAND Corporation, MG-595/5-OSD, 2008.

securing the people's trust and cooperation are bound to suffer if the government's forces or foreign forces acting on its behalf commit violence against the neighbors, friends, and family of that very population. This age-old lesson had to be relearned at great cost in Iraq and Afghanistan, where U.S. troops now employ sounder, gentler tactics in order to engender trust, not fear, on the part of the general public.[5]

When U.S. forces face the need or chance to destroy insurgents, they are usually left with a binary decision between deadly force and no force. This all-or-nothing choice reinforces the reluctance of senior commanders to authorize subordinates to use force at their discretion in sensitive circumstances (e.g., in urban areas) for fear that they may err with unfortunate repercussions. Such reluctance can consume precious time, squander opportunities, take decisions out of the hands of those directly involved, and heighten risks to both troops and their mission.[6] Among the most valuable qualities in COIN are local initiative, responsiveness, quickness, and flexibility. Yet, these qualities are sacrificed when command and control are centralized for fear of mistakes. In sum, even as U.S. forces are getting better at COIN, they are handicapped by a lack of capabilities and preparation to take action against insurgents without risking harm to the people they are supposed to protect.

Non-COIN Missions

Until this point, our rationale for a military continuum of force has been based mainly on COIN in the Muslim world. This is the most obvious case in which a full range of options appears advantageous, and it also may be the most important one, given the likelihood of future Islamist-based violence. However, there is a broader basis for

---

[5]   U.S. Army and U.S. Marine Corps, FM 3-24S/MCWP 3-335, *U.S. Army/Marine Corps Counterinsurgency Field Manual*, Washington, D.C., December 16, 2006.

[6]   The case for distributed authority and decision-making is best laid out in David S. Alberts and Richard E. Hayes, *Power to the Edge: Command . . . Control . . . in the Information Age*, Command and Control Research Program, U.S. Department of Defense, Washington, D.C., 2005.

this need that comes from other missions and other regions. It is useful for two reasons to consider these other needs in analyzing operating requirements. First, doing so can help determine how strong the general case is for investment in systems and training for such capabilities, given that both monetary and human resources are scare. Second, doing so can help ensure that requirements take into account as many missions and conditions as possible so that capabilities can be designed to be versatile.

Accordingly, this study looks at a wide variety of operations amid populations, including humanitarian intervention, peacekeeping, protecting U.S. officials and property against anti-American violence, supporting U.S. civil authorities confronted with domestic disorder, and maritime contingencies. This aperture should be wide enough to permit general continuum-of-force requirements to be identified.

Humanitarian Intervention

The United States, like its allies, has been reluctant to use its forces to stop mass killing and other atrocities. It did not do so in Rwanda (no one did), Sierra Leone (the UK did), or East Timor (Australia did). It did so in the Balkans, albeit belatedly and largely with air power. It did so in a physically tiny but symbolically significant way in Liberia. It has not done so in Darfur. As a rule, the United States acts to stop mass atrocities only when it has special interests or clear responsibilities to do so.[7] Nonetheless, U.S. (and allied) forces must be prepared for such missions, especially as the principle of responsibility to protect (R2P) gains political support.[8]

---

[7]   Whether specific instances of mass killing are called "genocide" or not does not appear to matter in U.S. decisions to intervene. Rwanda was indisputably genocide, but the United States declined to say so and chose not to intervene. The United States has called Darfur genocide, but again has chosen not to intervene. Therefore, the scale of killing is not a good predictor of whether U.S. forces will engage.

[8]   The R2P principle, adopted in 2005 by the UN Security Council and General Assembly, holds that the international community has the responsibility (and, by implication, the right) to intervene with whatever means are necessary when a government engages in or fails to prevent mass atrocities against its own people.

Many past and simmering cases of mass atrocities exist in sub-Saharan Africa. With growing interests there—oil, potential terrorist encroachment, Chinese inroads—the United States has created Africa Command (AFRICOM), a regional joint combatant command for Africa. This signifies more than a desire to become more aware of what is happening in Africa: The United States is likely to be cooperating more frequently and closely with African countries and the African Union as they seek to secure their regions.[9] U.S. forces already perform assistance and advisory functions, and it is possible that they will play either direct or enabling roles in actual military interventions to stop mass killing or other crimes against humanity. Even if the Africans were to provide most of the ground forces, the United States and its Western allies might have to provide special forces; logistics; command, control, and communications; air mobility; and other specialized capabilities that the Africans lack. Under extreme (Rwanda- or Darfur-like) conditions, and depending on U.S. interests, U.S. ground forces might be called on to participate in operations.

Intervention to stop mass killing is unlikely to involve force-on-force combat with regular military opposition. More often than not, such atrocities are committed by ragtag militia, child armies, gangs of armed civilians, or other irregular forces. Even when the state's own security services are involved, they operate off duty as death squads or undisciplined small units (with or without direction from high command). Cambodia, Bosnia, Rwanda, Sierra Leone, Darfur, East Timor, and most other killing fields have revealed that such killers are rarely also serious fighters. Their enthusiasm wanes when confronted with capable opposing troops. For this reason, it may not be necessary to kill the killers in large numbers. In Sierra Leone, for example, British paratroopers had only to eliminate a handful of killers before the entire 30,000-strong rebel force disintegrated.

Of course, sparing the lives of those engaged in genocide is not a constraint, with one notable exception: young boys, possibly coerced

---

[9] Apart from creating AFRICOM, the United States has increased greatly its foreign assistance to Africa and is concentrating most of its global peacekeeping capacity-building effort there.

into such service (and perhaps doped). Child soldiers are increasingly common in African civil wars and mass killings. They are considered to be generally less disciplined, stable, and fearful than older fighters and thus more dangerous.[10] Consequently, if regrettably, forces sent in to stop killing may feel particular urgency to fire when confronted by child soldiers, suggesting a strong case for a continuum of force. Child soldiers are already considered expendable by the warlords who snatch, drug, and arm them. Options are needed to stop them without taking their lives.

There are additional reasons why a continuum of force is important in this setting. First, it is not always clear who the killers are, unless they are spotted in the act. Second, more than one side may have a hand in the violence, in which case, what appears to be vicious killing may in fact be vicious self-defense. Third, eventual reconciliation may be more difficult if significant deadly action by U.S. forces on behalf of one side contributes to an urge for revenge.

Generally speaking, U.S. (or any other external) military intervention to stop mass atrocities would be more effective and less risky with than without a continuum of force. While this may not be sufficient to justify investment in capabilities to fill the current void, it could add to other justifications while also affecting requirements.

Peacekeeping[11]

Although it is not a major mission area, U.S. forces have been and likely will be engaged in peacekeeping.[12] In standard multilateral peacekeeping —which involves interposing international forces between, and with

---

[10] Author interview with UN Police Advisor, Police Division, UN Department of Peacekeeping, New York, October 11, 2007.

[11] This study has benefited from contact with United Nations Department for Peacekeeping Operations, the UN's peacekeeping organization, which has struggled with similar problems regarding missions amid populations that may be unfriendly or conceal dangerous individuals.

[12] Peacekeeping may or may not be conducted under UN auspices or operational control. For instance, the North Atlantic Treaty Organization (NATO) may take up peacekeeping responsibilities under a UN mandate but not UN control. Whatever the legal and institutional context, the issues bearing on the use of force are generally the same.

the consent of, warring parties—strict, agreed, and consistent rules of engagement (RoE) determine when and what force may be used. The tightest RoE permit peacekeepers to use force only for self-protection, and the weaponry provided (e.g., sidearms only) may reflect this. Beyond this, RoE may be specifically relaxed to permit forcible action to ensure the security of noncombatants or aid providers, respond to cease-fire violations, clear areas or routes, control access, or quell spontaneous violence. In addition, U.S. forces may be engaged in peace-enforcement operations under nonpermissive or semipermissive conditions in which not all parties are genuinely on board. Indeed, U.S. forces are more likely to participate in very challenging peace operations than in operations that forces from scores of other countries can handle. In the extreme, RoE might permit forcible action to bring intransigent factions or forces into consent and compliance.[13]

In such contexts, the availability of options between lethal force and inaction could be very beneficial. This is true for the most-restrictive and least-restrictive cases and, therefore, for everything in between. To illustrate, assume RoE permit deadly force only for self-protection. The premise in such a case is that the agreement to cease fighting and accept peacekeepers may be fragile. One or another side may abrogate this agreement if peacekeepers use deadly force. Even self-protection by nonlethal means could be advantageous. It could be even more advantageous to have nonlethal options to produce desired effects beyond self-protection (e.g., responding to violations or protecting citizens) if strict RoE apply.

Even if U.S. forces are infrequently engaged in peacekeeping, the United States has a strong interest in the success of such operations. This raises a question concerning the continuum of force that merits serious consideration: Should the United States share at least some of its nonlethal capabilities with others? If so, how widely (e.g., with NATO and the UN)? There is a strong case to do so. If the United States develops such capabilities for missions other than peacekeeping, it would presumably want to place such capabilities in the hands of countries

---

[13] At such a point, academic and diplomatic purists would say that the term *peacekeeping* no longer applies.

that could put them to good use in peacekeeping. A more definitive answer depends on the sensitivity of technologies used for these capabilities and the possibility that the technologies or capabilities could be used against U.S. interests.

In sum, whether for U.S. forces or for others, a continuum of effective force would represent major progress in peacekeeping. As it is, the use of lethal force may either exceed agreed RoE or risk triggering violence and a breakdown of peace. To be clear, there are circumstances in peace operations in which deadly force may be not only permissible and justified but also efficacious (e.g., to send a clear message of the consequences of violation or interference). However, this is not an argument against having options to produce effects without using deadly force. Again, peacekeeping requirements alone may not justify significant investments. But the advantages of more-effective peacekeeping, including by other nations, are worth taking into account in considering a continuum of force.

Protection of U.S. Personnel and Property

Populism, nationalism, socialism, antiglobalization, and anti-American sentiment could intensify and spread in, for example, Latin America.[14] If they do so, the probability will rise that U.S. interests could be endangered, with or without the connivance of Latin American regimes. Appropriation of commercial assets, disruption of commercial operations, state-run media agitation, and accusations of U.S. Central Intelligence Agency involvement are not without precedent in the region. Physical threats to U.S. government personnel and property, though unlikely to be widespread, could arise.

In such a future, the United States would have to act judiciously lest it aggravate conditions or at least stoke anti-Americanism. The very thought of U.S. military intervention in Latin America—against the historical backdrop of El Salvador, Cuba, Grenada, Panama, Nicaragua, and other campaigns—would be exploited by those with an anti-American agenda. This is not to suggest that the United States must

---

[14] While one naturally thinks of Venezuela, Ecuador, and Bolivia, such movements could also gain traction in Argentina, Central America, and even Mexico.

rule out armed intervention in Latin America under all circumstances. That is a policy question—a hypothetical one at that—that goes well beyond the purpose of this study. But the United States must be able to secure its legitimate interests in the Western hemisphere without being heavy-handed.

In this context, U.S. armed forces—perhaps embassy Marine guards, perhaps a larger force—could be called upon to protect a U.S. official presence from angry, swelling mobs. Perhaps local security forces would act promptly to prevent a violent confrontation, perhaps not. Beyond doubt is that modern media would cover and spread pictures and stories of events—accurately or otherwise—throughout Latin America, including into every city and village via satellite TV, and to every pocket of anti-Americanism and extremism via the Internet.

Thus, conditions could include a small U.S. force confronted by a large mob; a lack of support from (or unhelpful actions by) local security services; extreme political tension, local or hemispheric; intense and not necessarily fair news coverage; and a strong possibility of such events being emulated, possibly orchestrated, elsewhere.

Quelling Civil Disturbance

Americans are uncomfortable with U.S. military forces being employed domestically except under rare and tightly circumscribed conditions.[15] This is not just a public allergy: Its roots run deep in the country's institutions and Constitution. The national government's use of military power to enforce states' compliance with national law, as in the school desegregation confrontations of the mid-20th century, raises profound issues. Of course, any reliance on the use or threatened use of deadly force or even less-than-lethal violence only compounds the problem. The death or harm of U.S. citizens or other residents by the hand of the U.S. military is at best an alien notion and at worst a horrific thought for most Americans.

---

[15] Beyond the political and psychological inhibitions on domestic use of the armed forces, the legal doctrine of *posse comitatus* is firmly established in U.S. statutes. It sets high procedural hurdles, including the practice that state governors or other civil authorities must request support from the armed forces; these forces then function under tight civil control.

Yet, armed forces will almost certainly have to be used domestically again. Least controversial, of course, is the involvement of National Guard forces under gubernatorial control, or federal control in response to gubernatorial request, during natural-disaster relief operations. However, we now know that when a major hurricane slams into a U.S. city, a breakdown in law, order, and civility can occur. When the breakdown reaches a certain scale or severity, local and state law-enforcement capabilities may not be sufficient.

If innocent lives are threatened under such circumstances, and the use of force cannot be excluded, it is vastly better for police to make such decisions and take such actions, with the armed forces there to back them up as needed with support and for deterrence. Bad as it may be for U.S. military units to use deadly force inside the country against persons who are themselves using deadly force, it could be worse if they did so against persons engaged in criminal and unruly but not homicidal behavior. At the same time, military forces lacking nonlethal and scalable capabilities may be confronted with choices more unpalatable and risky than any they face abroad.

Such situations underscore the importance of at least some military forces being capable of undertaking law-enforcement services. To some extent, this is a matter of doctrine, preparation, and mind-set: Good police are conditioned to avoid or minimize violence, whereas military forces are conditioned to destroy enemy forces. The gulf that separates these two cultures is as wide as the difference between crime and combat. While this raises the matter of training U.S. military services to use force judiciously, it also suggests that the possession of scalable-effects capabilities could be of great value in domestic contingencies. Although this need pertains to the National Guard, active forces could benefit from similar preparations and capabilities. Just as peacekeeping missions suggest a need to share ideas and even capabilities with UN peacekeeping organizations, the possibility of domestic use of force argues for closer cooperation than now exists between U.S. military and police organizations.

Conclusions

From this brief survey, one is struck by the diversity of potential U.S. military missions other than traditional warfare and, thus, by the difficulty of generalizing. Lumping such missions together as "irregular operations" or "stabilization operations" risks inadequate preparation for missions that can differ as much from one another as they do from regular combat. Some border on police work—e.g., enforcing order and managing crowds—which can fall to military troops when police are unavailable or inadequate, which is often the case abroad. In other situations, including COIN, enemy fighters operate in urban areas to protect themselves or to endanger the people among whom they hide. Missions against enemy combatants typically entail different objectives, RoE, and tactics than those involving the control of noncombatants. Still, the common aspect of operating amid populations points toward a common need for options along the continuum of force.

This chapter's look at the policy setting and relevant military missions leads to several general conclusions. First, the growing likelihood and significance of operations amid populations suggests a need for U.S. military forces to be able to gain control of situations, carry out their tasks, and protect themselves without using deadly force. This is especially critical when the attitude of the populations in question is critical for success (as it is in COIN, peacekeeping, and humanitarian intervention) and when enemy fighters cannot be readily distinguished or separated from ordinary people, as is often the case. A primary strategy of terrorists, insurgents, and other U.S. adversaries is to hide in populations, in effect daring U.S. forces to attack them there and thus risk hurting civilians. Lacking a continuum of force, the United States is not fully capable of countering this strategy.

Although nonlethal options have long been essential in law-enforcement missions, in which restoring public order with minimum violence is stock-in-trade, they have been treated by the military as having only marginal utility in only rare circumstances. Although insurgents present dangers that exceed the capabilities of police, U.S. military forces could erase a major disadvantage in COIN and other important missions if they had a range of nonlethal options. Such options would offer small units and their commanders more flexibility, autonomy, and responsiveness; reduced risk of committing mistakes with wide ramifications; and better odds of accomplishing their missions.

Second, being able to disable persons without killing them is too low a standard. Acts of violence against populations whose trust and cooperation U.S. forces need to earn—and which are key to those forces' ultimate success—may undercut missions and entire campaigns. Harmful or painful nonlethal force may turn a crowd into a mob, a mob into a riot, a riot into a confrontation, or a confrontation into a cause célèbre for an insurgency. The ability to calibrate nonlethal effects from none to low, low to moderate, and moderate to high is as important as mere nonlethality.

Just as police do all they can to avoid harming innocent persons domestically, U.S. forces must have the means to carry out their duties without injuring, terrifying, or inducing animosity among the populations in which they and their enemies operate. Ultimately, the need for nonviolent effects lies in the fact that U.S. forces are often as responsible for the security of an indigenous population as they are for the elimination of the enemy fighters who hide in that population. True, U.S. forces must protect themselves—but not at the cost of harming the people they are supposed to defend.

This suggests a need for mild effects that can dissuade or disperse but not hurt noncombatants while giving U.S. forces the ability to seize the initiative over enemy combatants in dangerous situations. For example, having ways to cause fighters and innocent persons to respond differently or to separate into distinct groups would permit more-forceful action against the former while minimizing harm to the latter. Threading this needle may seem hard, but it is possible, as the next chapter demonstrates.

Third, the void between lethal action and inaction ought to be filled to the extent possible with a standard set of capabilities available to regular small units engaged in COIN, peacekeeping, humanitarian intervention, and other irregular operations. This general requirement lies in the fact that the U.S. military does not, for the most part, have specialized forces for such missions but instead uses the same force types for each. Moreover, the nature of these missions is such that these units may not know every morning the sorts of predicaments and persons they will face that day.

U.S. forces embarking on a mission may be able to foresee the need for and take along specialized nonlethal capabilities. Or, they may have the time and transportation options that allow them to call in specialized capabilities too scarce or bulky to carry themselves. However, the need for options ranging from nonviolent means to lethal force seems to be both frequent and unpredictable, and the need for a unit to respond without delay militates against keeping such capabilities at the brigade or higher level. Accordingly, we looked especially for versatile capabilities that a small unit could carry and use. These conditions also imply that the capabilities should be affordable enough that most small units can be equipped with them. Finally, it would be beneficial if such capabilities were scalable from nonviolent to violent but nonlethal and even to lethal effects.[16]

Fourth, the missions and conditions that give rise to the need for a continuum of force are not circumstances of the distant future: They are occurring now. As of this writing, U.S. troops find themselves in situations in which possessing nonlethal and nonviolent options could mean the difference between success and inaction, failure and U.S. casualties. All else being equal, solutions to the continuum-of-force void that are already at hand or only a few years off are preferable to those that are speculative and may require a decade of research. This argues for examining established technologies (or at least understood phenomena) over those based on unproven science.

In sum, a need exists for a versatile set of scalable capabilities that small military units can carry and use to gain advantage over enemy fighters without hurting, alienating, or killing people whose well-being U.S. forces are there to protect and whose cooperation U.S. forces need. Clearly, this requirement goes well beyond the simple need for weapons that can incapacitate their targets without killing them.

---

[16] The particular advantages of scalable-effects capabilities have been identified in existing military literature. See, for instance, Lamb, "Emerging Nonlethal Weapons Technology and Strategic Policy Implications for 21st Century Warfare," 2003.

CHAPTER THREE

Possibilities

Stretching Our Thinking

Before analyzing in detail the capabilities required for a continuum of force, there is a need to assess whether such a continuum is possible, broadly speaking. The basic standard of effectiveness for the continuum of force suggested here is the ability to control a situation or group of people without killing, harming, or alienating noncombatants while at the same time disadvantaging any enemy combatants in their midst. By *disadvantaging*, we mean restricting and reducing the ability of enemy combatants to carry out hostile intentions, interfere with U.S. missions, or fight another day. Meeting this standard will not be easy. To begin with, it will require expansive and creative thinking about the nature of the problem, how similar problems are dealt with outside the military, and how to harness technology.

Accordingly, this chapter offers three perspectives to illuminate ways to create a continuum of force:

- Solving the problem depends as much on managing information, time, and decision-making as on causing physical effects.
- For all the differences between military operations and law enforcement, it is critical to study how police seek to exert control, ensure order, and apprehend dangerous persons with minimal violence.
- With technology progressing on many fronts, the search for continuum-of-force capabilities must be open, broad, and ingenious.

Time, Knowledge, and Judgment Along the Continuum

If U.S. troops knew the composition, intentions, and capabilities of the foes they face during the missions and conditions described in Chapter Two, they would know what level and type of force, if any, to use against them in order to be confident of success. Usually, however, U.S. troops will not have such knowledge. Therefore, even with the capability to deliver a range of effects, U.S. troops will often be unsure of what effects to choose. The challenge of operating amid populations is as much one of knowledge as it is of having capabilities to cause desired physical effects. Knowledge—information, reasoning, problem-solving—makes a continuum of military force possible.

Recent analysis from the National Defense University suggests that the key to success in COIN and other complex operations is to maximize *time-information*—the product of useful information and the time needed to form reasoned judgments with that information.[1] With useful information increasingly available thanks to advanced sensors and networking, it should be possible to improve judgment in unclear and urgent conditions, such as those arising during COIN. One of the advantages of good decision-making under such conditions is that it can buy time, which in turn permits more information to be gathered. With more information and adequate time to process it, preliminary decisions can be validated, refined, or revised. This type of adaptive decision-making melds the virtues of experience-based intuition and objective analysis of new information.

What does this have to do with a continuum of force? As previously explained, today's insurgents often look like the people among whom they operate. Their willingness to sacrifice ordinary people, even members of their own community, is key to their strategy, as evidenced by their reliance on populations as shields and their wanton use of suicide bombing. Defeating such enemies without endangering the people around them is, as already noted, a severe challenge for U.S. forces.

---

[1]   David C. Gompert, Irving Lachow and Justin Perkins, *Battlewise: Seeking Time-Information Superiority in Networked Warfare*, Center for Technology and National Security Policy, National Defense University Press, Washington D.C., July 2006.

Both time-information and a continuum of force are important in meeting this challenge. Situations in which nonlethal or nonviolent measures may be indicated tend to be characterized by uncertainty and urgency. Consequently, increased time-information can improve judgments about the level and sort of force to use. In turn, using the right level and sort of force can create the opportunity for acquiring more time-information by, for example, causing innocent persons to disperse and determined enemy fighters to take a stand. If U.S. forces possess capabilities scalable from nonviolent to lethal effects, they can calibrate their actions according to the information and time they have *and* can use such capabilities to gain more information and time.

This demands advanced methods of collection, communication, and use of information about and from the population. Recent RAND research proposed the development of an integrated counterinsurgency operating network (ICON) to harness the power of information more fully and inclusively.[2] ICON is predicated on the research finding that most of the information needed by COIN forces can be collected through open communication with the population. Such a network could improve appreciably the timeliness and quality of decision-making and thus of troop performance in COIN.[3] Related work identified ways to hone cognitive abilities needed for COIN, such as by more-strategic recruitment, education, training, and decentralized command and control to stress distributed decision-making under urgent and complex circumstances.[4]

These improved information and cognitive capabilities would enable operating units to act against insurgents without harming noninsurgents. Successful COIN campaigns are the result of good awareness and judgment, not brute force. Indeed, the better the aware-

---

[2] Martin C. Libicki, David C. Gompert, David R. Frelinger, and Raymond Smith, *Byting Back—Regaining Information Superiority Against 21st-Century Insurgents: RAND Counterinsurgency Study—Volume 1*, Santa Monica, Calif.: RAND Corporation, MG-595/1-OSD, 2008.

[3] Libicki et al., *Byting Back*, 2008, p. 129.

[4] David C. Gompert, *Heads We Win: The Cognitive Side of Counterinsurgency (COIN): RAND Counterinsurgency Study—Paper 1*, Santa Monica, Calif.: RAND Corporation, OP-168-OSD, 2007.

ness and judgment, the less the need for sheer violence. Improved judgment and decisions made possible by information networking can reduce the risks of mistakes of using force in the ambiguous and sensitive conditions of COIN. However, when only lethal options are available, even more information and better decision-making cannot assure that the actions of U.S. forces will be effective when measured against the ultimate standard: convincing the population to reject insurgency and side with the government. Both improved time-information and a continuum of force are needed.

To illustrate, say that a small U.S. military unit has the mission of securing a densely populated neighborhood in which residents are ambivalent and insurgents are active. In one case, the unit's commander lacks knowledge and must act urgently. Even with nonlethal options, the commander's concern for the mission and troop safety may dictate the use deadly fire despite its risks to noncombatants. Having used such force, the commander may find that opportunity to acquire more time and information has vanished. In another case, assume that the commander is able to gain additional time and information by communicating with local authorities and the population. In this situation, the unit's actions could be more measured *if* a range of options short of deadly force is available. If they are, using them could buy both more time and more information, permitting an adjustment in the level of force, especially if the capabilities are scalable. Of course, the commander's cognitive abilities are critical to exploiting such an opportunity. The interrelated contributions of time, information, the force continuum, and decision-making are depicted schematically in Figure 3.1.

One of the most important benefits of such a combination is that it improves conditions for timely yet sound decision-making by those in the best position to decide: commanders on the scene. As John B. Alexander notes, "the pace at which operational situations can change is accelerating. Therefore, the authority to [use nonlethal weapons and transition to lethal weapons] will be pushed to lower and lower levels."[5]

---

[5] John B. Alexander, "Non-Lethal Weapons to Gain Relevancy in Future Conflicts," *National Defense*, March 2002, pp. 30–31.

Figure 3.1
The Interconnected Contributions of Time, Information, the Continuum of Force, and Decision-Making

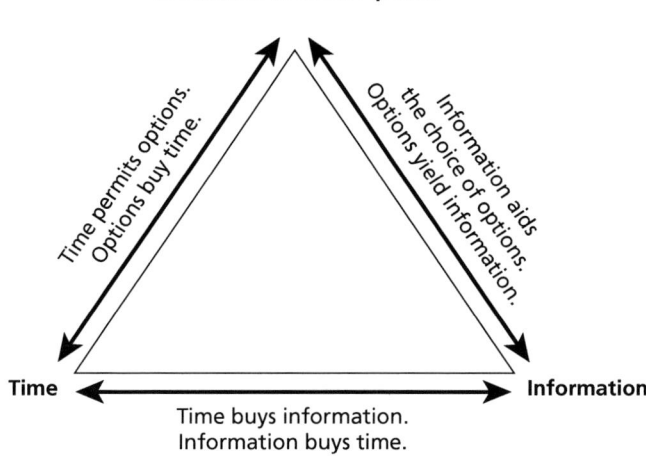

Continuum-of-Force Options

Time permits options.
Options buy time.

Information aids the choice of options.
Options yield information.

Time

Information

Time buys information.
Information buys time.

The availability of a continuum of force should encourage such delegation of authority, given the reduced risk of civilian death or harm.

As it is, risks associated with the choice to use deadly force reinforce habits of centralized decision-making that hamstring COIN operations. When senior commanders hoard the authority to use force, they discard the considerable but fleeting value of "tacit knowledge" available only to junior officers in the field.[6] This practice devours precious time without yielding more information. Greater latitude permits more-informed, timely, and responsive decisions, and nonlethal options permit greater latitude.

According to one of its practitioners, "counter-insurgency is a thinking man's sport."[7] Especially against the likes of today's Islamist insurgents, it will be won or lost more by brainpower than by firepower. Broadly speaking, the cognitive demands of COIN, like its informa-

---

[6] Gompert, Lachow, and Perkins, *Battlewise*, 2006, pp. 34–36.

[7] Colonel Jim Pasquarette, U.S. Army, Iraq, quoted in David Ignatius, "Fighting Smarter in Iraq," *Washington Post*, March 17, 2006, p. A19.

tion demands, are more complex than those of regular war. One of the effects of the void between lethal action and inaction is that it fails to exploit and can even hinder cognitive performance.

The possibility of significant enhancements in time-information and decision-making despite uncertainty and urgency is already at hand. Sensing and networking technology have made great advances in the last decade or so, and the military is developing training techniques that will permit its forces to make better sense and use of information during operations.[8] These developments offer forces the opportunity to use nonlethal and nonviolent capabilities more effectively and with less risk. ICON would enhance a continuum of force.

Nonmilitary Experience

A continuum of force requires not only technical and cognitive capabilities but also a set of sound practices and standards for training and actual use. The U.S. military is devoting more attention to this need.[9] But police forces, particularly large metropolitan services that have faced serious public-order challenges and threatening actors in dense populations, have extensive and rich experience with developing and implementing such capabilities, practices, and standards.

While military troops are conditioned to destroy enemy forces, police are conditioned to minimize violence. The latter operate within populations that count on them to protect and certainly not hurt them. In legitimate states, the authority of police to use force comes, in effect, from the population.[10] This basic understanding places a responsibility on police to make the population's safety their foremost consideration. Consequently, police are trained to use deadly force as a last resort. The conditions in which they may use force and the amount of force they

---

[8]   Gompert, Lachow, and Perkins, *Battlewise*, 2006, pp. 24–26.

[9]   U.S. Army, U.S. Marine Corps, U.S. Navy, and U.S. Air Force, *NLW Multi-Service Tactics, Techniques, and Procedures for the Tactical Employment of Nonlethal Weapons*, 2007.

[10]   New York City Police Department Police Academy, *City of New York Police Department Police Student's Guide: Use of Force*, New York, July 2005.

may use are tightly restricted. In the New York City Police Department (NYPD), for example, deadly force is sanctioned in advance only in the face of an imminent threat of deadly force being used against police or citizens.[11] Of course, in those rare cases when military forces are called upon to keep order within the United States, deadly force is sure to produce national outrage.[12]

Police departments have established force-escalation standards that essentially constitute a use-of-force continuum. With some variations, such standards are common across departments in the United States. Table 3.1 is a model developed by the Florida Department of Law Enforcement Criminal Justice Standards and Training Commission Defensive Tactics Task Force. It shows that officers are expected to

Table 3.1
Police Use-of-Force Scale

Level	Subject Resistance Type	Officer Control Options
One	Presence; psychological intimidation	Presence; field interrogation stance; preemptive physical control
Two	Verbal resistance; nonverbal clues	Dialogue; verbal direction; touch; preemptive physical control; further escalation
Three	Passive physical resistance	Transporters; pain compliance; takedowns; restraint devices; counter moves
Four	Active physical resistance	Intermediate weapons
Five	Aggressive physical resistance	Techniques to temporarily incapacitate subject
Six	Aggravated physical resistance	Lethal defensive action

SOURCE: Florida Department of Law Enforcement, Criminal Justice Standards and Training Commission, Defensive Tactics Curriculum, "Legal and Medical Risk Summary," June 2002, pp. 1–9.

---

[11] New York City Police Department Police Academy, *City of New York Police Department Police Student's Guide*, 2005.

[12] As it did in 1968, when National Guardsman killed several students at Kent State University.

start with the lowest reasonable level of force and escalate to higher levels based on increased resistance from and knowledge of the subject.

Officers are expected to consider the following factors when making decisions regarding the appropriate level of force. Some factors refer to the subject, others to the police officer:

- Subject factors
  - seriousness of the crime committed by the subject
  - size, age, and weight of the subject
  - apparent physical ability of the subject
  - subject's medical conditions and mental state, and the influence of alcohol or drugs
  - number of subjects present who are involved or may become involved
  - weapons possessed by or available to the subject
  - presence of innocent or other potential victims in the area
  - whether the subject can be recaptured at a later time
  - whether evidence is likely to be destroyed.
- Police factors
  - size, physical ability, and defensive-tactics expertise of the officer
  - number of officers present or available
  - the necessity for immediate reaction in the case of sudden attack
  - weapons or restraint devices available to the officer
- legal requirements and department policy.[13]

NYPD employs the standards listed in Table 3.2.

Note that deadly force is to be used by police, according to these standards, only when they or the people they are obliged to protect face an imminent threat of death or serious injury. Even at that point, the purpose of deadly force is lethal *defensive* action. Only in cases of extreme and unambiguous danger to public safety may law enforce-

---

[13] University of Florida Police Department, *Department Standards Directive: Use of Force*, March 2007.

Table 3.2
NYPD Escalating Scale of Force

Provocation or Condition	Appropriate Force Response
Imminent threat of death or serious physical injury	Deadly force
Threatened or potential lethal assault	Drawn or displayed firearm
Physical assault; threatened or potential physical assault likely to cause physical injuries	Impact techniques; pepper spray
Minor physical resistance	Compliance holds
Verbal resistance	Physical force
Refusal to comply	Command voice
Minor violations or disorderly conditions	Verbal persuasion
Orderly public places	Professional presence

SOURCE: New York City Police Department Police Academy, *City of New York Police Department Police Student's Guide*, 2005.

ment actions begin to resemble military combat, in which the destruction of the "enemy" is the only way to eliminate the threat. Even then, the paramount consideration is the safety of the population.

While it is rare for police to find themselves in combat situations, it is less and less rare for U.S. military forces to be called upon to operate in conditions like those described in the first column of Table 3.2. Again, such situations may arise out of military missions such as those described in Chapter Two; alternatively, U.S. military forces may have to provide law and order in situations abroad because of the lack of adequate police.[14] Thus, the relevance to military operations of police doctrine and standards regarding use of force lies less in the possibility that police must engage in combat and more in the virtual certainty

---

[14] According to UN officials, the number of professional police available, with some warning and preparation, to deploy into difficult situations worldwide is only 17,000 on paper and 12,000 in reality—a tiny fraction of the potential requirement for international security forces. Unless indigenous police are of sufficient quality and in sufficient numbers—which is often not the case—most of the total security requirement in conflict and postconflict situations must be met by military forces, be they UN, U.S., NATO, or other.

that military forces must conduct missions in which police doctrine and standards may apply.

At the same time, U.S. military forces operating amid foreign populations do not face the same formal restrictions as do police operating amidst the U.S. population. But the sensitivities may be as great, and the responsibilities may not be fundamentally different.[15] Again, when U.S. military forces are charged with providing security in another country, there should be no presumption that harming innocent unarmed foreigners is more permissible than harming innocent unarmed inhabitants of the United States. Just as U.S. security forces have obligations to U.S. citizens, they have obligations to those populations whose security is entrusted to them, whether by agreement, under UN mandate, or in accordance with some other politico-legal arrangement. Generally speaking, for missions other than regular combat against identifiable enemy forces—especially amid a vulnerable population—military forces should adhere to the principle that the population's safety is paramount. Therefore, to the extent that such missions and operations will continue to be common, military forces must be able to (1) minimize violence while carrying out their tasks and (2) avoid causing death, pain, or harm to innocent people.

We do not claim that the practices and standards developed by police departments should simply be adopted by the U.S. military, for missions and conditions differ too much to presume that can be done. Nor are the specific capabilities used by police necessarily the right ones for military forces, given that violent criminals and enemy fighters are different in kind. Still, what this excursion into the law-enforcement realm reveals is that there is a set of workable principles, practices, and standards that span wide ranges of situations and prepare law-enforcement forces to respond flexibly, purposefully, and carefully, despite uncertainty, to threats amid populations. This important aspect of a continuum of force for U.S. military troops is thus also possible.

---

[15] This assertion is borne out by the fact that status-of-forces agreements (SOFAs) often contain stipulations that require U.S. forces to observe the same restraints as the forces of the sovereign or require the approval of the sovereign for operations involving the use of force. The Iraq SOFA concluded on December 4, 2008, contains such a proscription.

Technological Possibilities

We deferred analysis of technologies needed for a continuum of military force until we identified requirements to inform that analysis. (Accordingly, requirements are detailed in Chapter Four and technological options in Chapter Five.) Nevertheless, it is worth noting in general how today's dynamic technologies could yield better solutions to the problem of operating amid populations. Indeed, many extant nonlethal weapons are based on decades-old (sometimes centuries-old) technologies, which are not up to the task in today's security environment.[16]

Progress in basic and applied science—e.g., waves, fields, sensing, dense energy storage, materials, miniaturization, precision delivery, biochemistry, physiology, and psychology—is expanding the potential for a continuum of force. New information capabilities improve opportunities to refine, calibrate, and aim force, as well as to sense and adjust to the effects of using force. It is precisely because technological options are less and less constraining that a continuum of force can be driven by strategy, policy, and operating needs rather than preconceived technical solutions, incremental thinking, or programmatic inertia.

In this regard, it is important to progress beyond a definition of nonlethal force that allows for shock, pain, and injury short of death. A more useful standard is whether the people exposed to the actions of U.S. troops will be less inclined subsequently to cooperate with those troops. Even if they are not deadly, projectiles, blunt-contact instruments, ingestible substances, and painful or disabling shocks will be deemed violent by persons on the receiving end; they may also be viewed as unjustified if those persons meant no harm to the forces that used these weapons against them. The possibility of injuries being captured on film, in cell-phone snapshots, or on satellite TV places a premium on capabilities that are less easily depicted as cruel or brutal.

This requirement suggests a need for capabilities that warn, disorient, dissuade, slow, or disperse people but *do not* injure them. It also underscores the need for capabilities whose scalability (i.e., ability to

---

16 John B. Alexander, "An Overview of the Future of Non-Lethal Weapons," *Medicine, Conflict and Survival*, No. 17, Vol. 3, July 2001, p. 183.

generate effects from the very mild to the severe) permits innocent persons to be separated from dangerous persons, who can then be disabled or killed. Such capabilities would take advantage of the opportunities presented in this chapter's two preceding sections: enhancing information, time, and adaptive decision-making techniques; and developing standards and practices along a scale of mild to lethal effects.

Certain capabilities may be useful only in specific circumstances. However, this study seeks those that are broadly useful. In this regard, directed-energy is a family of scalable technologies worth exploring insofar as such technologies can be used in a variety of conditions and missions to gain control without necessarily harming persons exposed to them. Directed energy has been associated with exotic, if not futuristic, weapons—e.g., anti–ballistic-missile and anti-satellite systems—but can also be applied at low levels of power to reduce damage. Light and sound, for example, are commonly used to warn or illuminate people without hurting them. This begs the question, to which we will return, of whether a continuum of directed energy could to some extent fill the gap between extremely mild and lethal force. Finally, we wish once again to mention that, given the importance of information about, observation of, and communication with the population amid which operations are conducted, progress in cellular telephony, personal-identification means, and video systems offers important opportunities.

Taking Stock

What, then, can be said at this point about the prospects for a continuum of force? Figure 3.2 shows a progression from binary to continuous effects, with each stage more difficult to attain.

Preceding chapters have explained the fundamental benefit of progressing from left to right in this figure, a movement that signifies U.S. forces' increasing capability to operate in urban areas to gain advantage over enemy fighters without harming ordinary people. To make such progress, U.S. troops need three things:

Figure 3.2
Binary-to-Continuous Effects

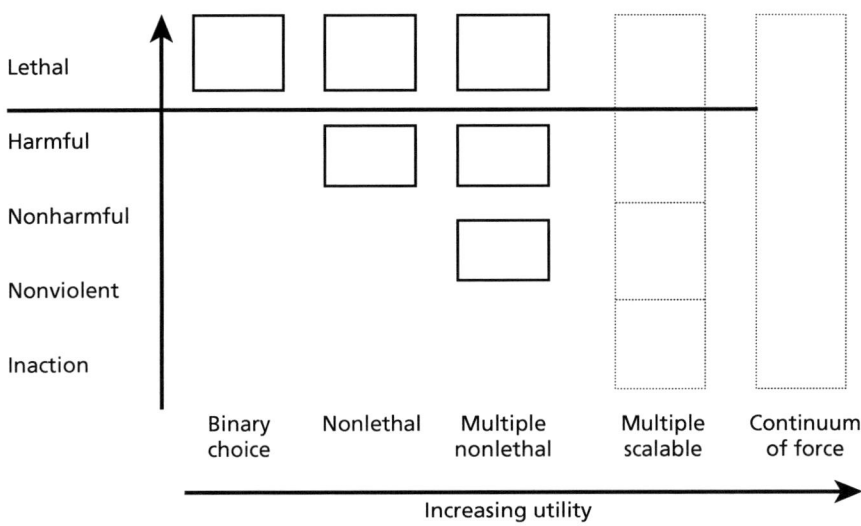

RAND *MG848-3.2*

- the ability to gain knowledge and time and to exploit both in adaptive decision-making under uncertainty and urgency
- standards and an understanding of how to escalate in order to gain advantage while managing risk
- scalable technologies that permit calibrated effects, from nonviolence to lethality.

This chapter has suggested that these basic conditions can be met, thanks to advances in the management of information; germane experience, principles, and practices from law enforcement; and progress in a wide assortment of potentially relevant technologies. With the general need for and possibility of a military force continuum established, the rest of this book examines requirements, options, a possible solution, and a path forward.

CHAPTER FOUR

Requirements

Developing Requirements

Chapter Two laid out a general case for making U.S. military forces capable of employing a continuum of force. Chapter Three indicated that such a capability ought to be possible. The next step is to identify the properties and composition, predicated on operating requirements, of continuum-of-force capabilities. Our method of doing so was to choose, describe, and analyze a large set of representative situations across a wide range of mission areas in which the binary choice between deadly force and inaction may be inadequate.

The chosen scenarios, presented by overall mission area, are

- *Eliminating important targets.* Specifically,
  - conducting deep urban penetration against low-value targets
  - conducting deep urban penetration against high-value targets (HVTs)
  - eliminating terrorists attacking participants in a religious pilgrimage
  - eliminating insurgents holed up in a mosque.
- *Providing public safety, order, and law enforcement.* Specifically,
  - confronting an angry mob with unclear intentions
  - rescuing hostages rapidly
  - apprehending a "most-wanted" charismatic cleric
  - searching a house for weapons.
- *Sealing off areas, cities, borders, and coasts.* Specifically,
  - confronting a challenge to a permanent checkpoint

- – confronting a challenge to a temporary checkpoint
  - – cordoning off a neighborhood
  - – ensuring port security
  - – securing an airfield.
- *Preventing mass atrocities.* Specifically,
  - – confronting child soldiers
  - – stopping citizen-on-citizen violence.
- *Conducting peacekeeping.* Specifically,
  - – operating within restrictive RoE
  - – patrolling an urban neighborhood in support of local police.
- *Protecting U.S. officials and property.* Specifically,
  - – resisting large-scale mob attack
  - – ensuring convoy security.
- *Supporting domestic civil authorities.* Specifically,
  - – preventing large-scale looting following natural disaster
  - – preventing unauthorized border crossings.

For each scenario, we analyzed the circumstances in which nonlethal or nonviolent force might be indicated and examined the consequences of having limited or no such capability. This analysis is summarized in Table 4.1.

These scenarios are important in two ways. First, they permit construction of a general framework for analyzing requirements for a military continuum of force, which is useful not only for these scenarios but for any military operations against threats amid populations. Second, they permit some observations about requirements. The remainder of this chapter deals first with the general framework and then with the requirements indicated by the scenarios.

Table 4.1
Summary of Scenarios

Scenario	Circumstances	Availability of Time	Persons Being Confronted	Objective(s) of U.S. Forces	Uncertainties About Circumstances and People	Consequences/Implications of Deadly Force
Eliminating Important Targets						
Conducting deep urban penetration against low-value targets	U.S. forces in a COIN campaign conduct (1) surveillance of low-level insurgent activity in a congested urban area and (2) a subsequent operation to seize insurgents in two identified safe houses.	Surveillance will occur over several days, but once relevant information is gathered, an ensuing operation becomes time-sensitive, requiring swift action within hours.	(1) Local civilians who inadvertently discover the surveillance efforts. (2) Low-level insurgents in safe houses. (3) Innocent civilians—women and children—collocated in safe houses. (4) A crowd of possibly hostile civilians who are in the vicinity of the operation.	(1) Gather intelligence on the identities of the insurgents; the location of the safe house; the insurgents' level of armaments and weapons training; and the insurgents' movement patterns. (2) Conduct operations to seize or eliminate insurgents based on intelligence. (3) Conduct operations that minimize the use of force against possible interference from local civilians; civilian casualties; and alienation of the population.	(1) The credibility of intelligence will determine the degree of certainty about insurgent numbers, location, movement patterns, and armaments. (2) The number of innocent civilians—women and children—collocated in safe houses who could be used as hostages or human shields. (3) The response by local civilians in the vicinity of operations: Crowd response could range from curiosity to hostility and could impede the quick exit of forces and the extraction of insurgents. (4) The ability of U.S. forces to differentiate combatants from innocent civilians if insurgents evade capture and blend in with the local population.	(1) Increased probability of civilian injuries or deaths. (2) Possible alienation of the population.

Table 4.1—Continued

Scenario	Circumstances	Availability of Time	Persons Being Confronted	Objective(s) of U.S. Forces	Uncertainties About Circumstances and People	Consequences/ Implications of Deadly Force
Conducting deep urban penetration against HVTs	U.S. forces assisting local security forces in a COIN campaign conduct a raid deep into an urban area to capture or eliminate two key insurgent leaders deemed HVTs. The insurgent leaders are meeting to discuss strategy in a 10-story apartment building.	There is small and critical window of time (approximately 30 minutes) for the execution of this operation: U.S. forces are compelled to act quickly to eliminate or capture the HVTs before they escape.	(1) Insurgent leaders and 15–30 armed guards. (2) A large number of civilians in the apartment building and surrounding city blocks.	(1) Capture or eliminate insurgent leaders and inflict a major blow to the insurgency. (2) Conduct operations that minimize the use of force against possible hostility from local civilians; civilian casualties; and physical destruction of the apartment building.	(1) Innocent civilians—women and children—collocated in the apartment building could be used as hostages or human shields. (2) The loyalty or allegiance of local civilians in the apartment building to the HVTs: If insurgent leaders escape, sympathetic civilians could offer refuge. (3) A number of civilians could be armed and impede the quick exit of forces and the extraction of insurgents. (4) The ability of U.S. forces to differentiate combatants from innocent civilians if insurgents evade capture and blend in with the local population.	(1) Increased probability of civilian injuries or deaths. (2) Possible alienation of the population if lethal force destroys the apartment building and kills scores or hundreds of civilians.

Table 4.1—Continued

Scenario	Circumstances	Availability of Time	Persons Being Confronted	Objective(s) of U.S. Forces	Uncertainties About Circumstances and People	Consequences/ Implications of Deadly Force
Eliminating terrorists attacking participants in a religious pilgrimage	U.S. forces in a COIN campaign seek to act on intelligence that terrorists (suicide bombers, most likely) plan to attack a religious pilgrimage of tens of thousands of pilgrims. The pilgrimage is occurring in an area (1) governed by only limited government control and (2) populated by a religious sect other than the sect that is on the pilgrimage. Intelligence suggests a coordinated attack at a chokepoint.	Should an attack occur, a rapid response is critical, and U.S. forces will have limited time—likely only minutes—to eliminate the terrorists and take control of the chaotic aftermath.	(1) A male or female terrorist (a suicide bomber, most likely) hidden amongst pilgrims. (2) Terrorists who may use vehicle-borne explosives along the route of the pilgrimage (at a chokepoint, most likely) to inflict maximum casualties.	(1) Eliminate the suicide bomber through sniper fire and prevent detonation of the suicide vest. (2) Calm a panicked crowd attempting to flee the site and instruct the crowd to get down in order to clear a field of fire for the sniper. (3) Reduce the number of casualties resulting from the use of lethal U.S. force, the detonation of a terrorist explosive, or both. (4) Prevent an attack that could further inflame sectarian hatred and civil strife.	(1) The level of government control in areas along the pilgrimage route. (2) Exactly where or when the terrorist attack will take place: Intelligence has not provided this information. (3) The ability of U.S. forces to differentiate the terrorist from innocent civilians if the terrorist blends in with the pilgrims or if a male terrorist disguises himself as a female to avoid being searched. (4) Cultural considerations (e.g., women are not searched because there are no female security forces at checkpoints) increase the likelihood that the terrorist could slip through checkpoints. (5) Whether the crowd would respond to instructions in the manner desired by U.S. forces.	(1) Increased probability of civilian injuries or deaths. (2) Further provocation of sectarian hatred and possible alienation of the religious sect if lethal force is used and perceived as being applied with disregard for the religious pilgrimage.

Table 4.1—Continued

Scenario	Circumstances	Availability of Time	Persons Being Confronted	Objective(s) of U.S. Forces	Uncertainties About Circumstances and People	Consequences/ Implications of Deadly Force
Eliminating insurgents holed up in a mosque	Local security forces have requested the assistance of U.S. forces to end a week-long siege by foreign militants who have taken control of a mosque and madrasah complex and hold hostages. Hundreds of students sympathetic to the militants have surrendered and provided information on the situation.	Time is an urgent factor in the conduct of any operation to end the siege: U.S. forces will have to launch an assault, incapacitate insurgents, and rescue hostages within minutes.	(1) Militants inside the mosque and madrasah complex. (2) Any students who have remained in the complex.	(1) End the siege and rescue the hostages with minimal casualties. (2) Conduct an assault that limits damage to the mosque, a historic national treasure. (3) Capture the militant leaders alive for interrogation purposes.	(1) The number and precise location of hostages who remain in the mosque. (2) The level of physical damage to the mosque by U.S. forces and the reaction of the population to such damage.	(1) Increased probability of civilian injuries or deaths. (2) Physical damage to the mosque could lead to extremist backlash and push moderates throughout the country to sympathize with an Islamic revolution. (3) Potential diplomatic and political disaster.

Table 4.1—Continued

Providing Public Safety, Order, and Law Enforcement

Scenario	Circumstances	Availability of Time	Persons Being Confronted	Objective(s) of U.S. Forces	Uncertainties About Circumstances and People	Consequences/ Implications of Deadly Force
Confronting an angry mob with unclear intentions	As part of a UN peacekeeping mission, U.S. MPs search a house that may contain illegal weapons and apprehend a suspect. As the operation unfolds, an angry crowd attempts to block the evacuation of the suspect.	The scenario is time-critical: The MPs need to respond quickly to the hostile crowd and, if needed, request reinforcements from a quick-reaction force within minutes in order to prevent the potential escalation of crowd violence.	(1) The suspect inside the house. (2) A crowd that has become aggressive toward the MPs.	(1) Conduct an unannounced search to apprehend the suspect and minimize the chance that illegal weapons are removed from the house. (2) Control the angry crowd and avoid using lethal force that could have serious negative repercussions on the peacekeeping mission and violate the RoE. (3) Minimize civilian injuries and casualties.	(1) The intentions of the crowd. (2) The escalation and level of the crowd's hostility and the potential for violence. These could impede the quick exit of the MPs.	(1) Increased probability of civilian injuries or deaths. (2) A possible resumption of violence in an area that has gained relative stability. (3) Potential diplomatic and political setbacks if relations with the local population are damaged.

Table 4.1—Continued

Scenario	Circumstances	Availability of Time	Persons Being Confronted	Objective(s) of U.S. Forces	Uncertainties About Circumstances and People	Consequences/ Implications of Deadly Force
Rescuing hostages rapidly	Terrorists have seized a local primary school and taken more than 200 children and a number of teachers hostage. Soon after the siege, six teachers are murdered and the terrorists threaten to kill the remaining hostages unless their demands are met. Local police are waiting for reinforcements and have requested immediate, rapid hostage-rescue assistance from U.S. forces from a nearby base.	Time is an urgent factor: U.S. forces must take action prior to the expiration of the terrorists' deadline. The forces will likely have only minutes to execute an assault, incapacitate terrorists, and rescue hostages with minimal casualties.	(1) Terrorists inside the primary school. (2) A large crowd of frantic relatives of the hostages.	(1) End the siege and rescue the hostages with minimal civilian injuries and fatalities. (2) Prevent the increasingly frantic crowd from breaking through the police barricade and entering the school.	(1) The exact number and precise location of terrorists and hostages in the school. (2) The length of time it will take for local police reinforcements to arrive. (3) The ability of local forces to calm the crowd and prevent any frenzied attempt to enter the school.	(1) Increased probability of civilian injuries or deaths. (2) Potential diplomatic and political disaster if the U.S. rescue operation fails or results in numerous casualties.

Table 4.1—Continued

Scenario	Circumstances	Availability of Time	Persons Being Confronted	Objective(s) of U.S. Forces	Uncertainties About Circumstances and People	Consequences/Implications of Deadly Force
Apprehending a "most-wanted" charismatic cleric	U.S. forces assisting a host nation in a COIN campaign conduct an operation to apprehend and arrest a popular cleric at his residence in a large town of approximately 50,000 residents. Intelligence indicates that the cleric is a key leader in the insurgency.	To maintain the element of surprise and minimize lethal confrontation posed by militias or sympathetic neighbors, U.S. forces will have to execute a precise raid, capture the cleric, and extract themselves within a short time (about 15–30 minutes).	(1) The cleric, his wife and five children, and 5–10 armed guards. (2) Civilians in the immediate vicinity of the cleric's home.	(1) Arrest the cleric without causing physical harm or injury to his family; inflict a major blow to the insurgency. (2) Conduct an operation that minimizes the use of force against possibly hostile local residents who support the cleric; minimize resulting civilian casualties.	(1) The response from local civilians in the vicinity of operations: U.S. forces are likely to come under fire from neighboring houses inhabited by civilians sympathetic to the cleric. (2) The number of innocent women and children in nearby houses.	(1) Increased probability of civilian injuries or deaths. (2) Further provocation of sectarian or religious hatred and overall alienation of the population.

Table 4.1—Continued

Scenario	Circumstances	Availability of Time	Persons Being Confronted	Objective(s) of U.S. Forces	Uncertainties About Circumstances and People	Consequences/ Implications of Deadly Force
Searching a house for weapons	U.S. forces in a COIN campaign conduct house-to-house search operations during daylight hours to uncover militia weapons in an urban neighborhood. RoE are restrictive, so U.S. forces use "cordon-and knock" tactics, which are less aggressive than "cordon-and-search" operations, to build rapport, legitimacy, and credibility with the local community.	There is more time available to conduct this operation, but U.S. forces are cognizant of the possibility of an ambush. Any such outcome would require a rapid response.	(1) Militia forces attempting a possible ambush. (2) Local civilians—including women and children—who may be hostile combatants.	(1) Determine the location of militia weapons caches and other potential threats. (2) Minimize the use of indiscriminate fire and civilian casualties in the event of an ambush or hostilities. (3) Conduct the search in a manner that builds rapport, legitimacy, and credibility with the local community and leads to relevant information.	(1) The location of militia weapons caches and the threat of potential ambush. (2) The response of local civilians, which may range from hostile to obstructive to neutral to aide to ally.	(1) Increased probability of civilian injuries or deaths. (2) Undermined U.S. legitimacy and credibility within the local community, which damages both rapport and a source of potentially valuable information.

Table 4.1—Continued

Scenario	Circumstances	Availability of Time	Persons Being Confronted	Objective(s) of U.S. Forces	Uncertainties About Circumstances and People	Consequences/ Implications of Deadly Force
Sealing Off Areas, Cities, Borders, and Coasts						
Confronting a challenge to a permanent checkpoint	U.S. forces in a COIN campaign are manning permanent checkpoints at key locations. A suicide bomber in a vehicle attempts to pass through a permanent checkpoint at a major U.S. forward operating base. The bomber is holding the driver of the vehicle and a woman and two children at gunpoint.	The available time for U.S. forces to respond is short—nearly nonexistent—due to the volatility and potential lethality of the situation.	(1) The suicide bomber in the vehicle. (2) Innocent civilians held at gunpoint in the vehicle and possibly used as human shields.	(1) Secure the checkpoint, prevent the vehicle from entering the base, and prevent the terrorist from detonating the suicide vest. (2) Use force that minimizes the injury or death of the innocent vehicle occupants.	(1) The rapid escalation of the situation and the possibility that suicide bomber will detonate explosives at the checkpoint. (2) A frenzied attempt by the woman and children to exit the vehicle: Sudden movements could lead to a split-second decision to use force.	(1) Increased probability of civilian injuries or deaths. (2) Possible alienation of the population.

Table 4.1—Continued

Scenario	Circumstances	Availability of Time	Persons Being Confronted	Objective(s) of U.S. Forces	Uncertainties About Circumstances and People	Consequences/ Implications of Deadly Force
Confronting a challenge to a temporary checkpoint	U.S. forces in a COIN campaign have set up a temporary TCP in a neighborhood experiencing a recent upsurge in violence. An approaching vehicle suddenly begins speeding in a different direction to bypass the TCP.	Time is less critical because the threat to the TCP is not immediate. However, quickly determining the potential threat posed by the fleeing vehicle is critical.	(1) Occupants in the vehicle who are attempting to avoid the TCP.	(1) Disable the vehicle and determine the threat it poses. (2) Use force that minimizes the injury or death of any innocent vehicle occupants.	(1) Whether occupants in the vehicle are combatants or innocent civilians. (2) The reason why the vehicle is bypassing the TCP and the potential threat that the vehicle and its occupants pose.	(1) Increased probability of civilian injuries or deaths. (2) Possible alienation of the population.

Table 4.1—Continued

Scenario	Circumstances	Availability of Time	Persons Being Confronted	Objective(s) of U.S. Forces	Uncertainties About Circumstances and People	Consequences/ Implications of Deadly Force
Cordoning off a neighborhood	U.S. forces in a COIN campaign are asked by the local government to deploy a battalion to cordon off a town of approximately 50,000 inhabitants and retake the insurgent-held neighborhoods. Checkpoints set up as part of the operation have caused considerable traffic and long lines more than 300 meters from the checkpoint itself. A speeding vehicle jumps out of a line and speeds forward.	This is a time-critical situation that requires a near-immediate response (i.e., within seconds): U.S. forces must halt the approaching vehicle and prevent a suicide attack.	(1) An insurgent with explosives on a suicide mission. (2) Local civilians in traffic and lines at the checkpoint.	(1) Stop the approaching vehicle and determine the threat it may pose. (2) Reduce the number of casualties resulting from the use of lethal U.S. force, the detonation of vehicle explosives, or both. (3) Calm a possibly dazed, angry, or hostile crowd in the confusion caused by the insurgent attack.	(1) The extent of the threat and the scale of the possible explosion. (2) The crowd's response to the explosion. Reactions could range from confusion to panic to anger to hostility toward U.S. forces.	(1) Increased probability of civilian injuries or deaths. (2) Possible alienation of the population.

Table 4.1—Continued

Scenario	Circumstances	Availability of Time	Persons Being Confronted	Objective(s) of U.S. Forces	Uncertainties About Circumstances and People	Consequences/ Implications of Deadly Force
Conducting maritime interdiction operations or opposed boardings	The U.S. Coast Guard conducts boardings of suspicious vessels in the waters off the southern coast of the United States. A fast-moving boat suspected of involvement in cocaine smuggling is intercepted.	After the interception, there is small and critical window of time in which to conduct the boarding. Operational decisions will become increasingly time-sensitive as resistance to the boarding increases.	(1) Armed drug smugglers.	(1) Intercept and board the vessel with a minimum use of force. (2) Conduct operations in accordance with U.S. and international maritime law.	(1) The number of smugglers on the vessel and the level of resistance to boarding.	(1) Increased probability of civilian injuries or deaths. (2) Potential diplomatic and political fallout.

Table 4.1—Continued

Scenario	Circumstances	Availability of Time	Persons Being Confronted	Objective(s) of U.S. Forces	Uncertainties About Circumstances and People	Consequences/ Implications of Deadly Force
Ensuring port security	The U.S. Coast Guard provides a security zone for an LNG tanker during its transit from the open ocean into the Boston Harbor. All marine traffic in the harbor is closed for these transits, which occur every few weeks and are well publicized. As the tanker moves through the harbor, an explosive-laden vessel speeds toward the tanker.	This scenario is very time-critical: Only minutes are available to halt the vessel and prevent a massive explosion.	(1) Likely terrorists on a suicide mission.	(1) Intercept and stop the vessel or the terrorists before explosives are detonated. (2) Minimize the use of force to decrease the risk of civilian casualties due to the unintended explosion of the tanker or suicide vessel close to the harbor.	(1) A narrow harbor and the tanker's proximity to civilians and businesses increases the risk that machine-gun rounds will kill or injure innocent civilians, unintentionally hit the tanker and cause an explosion, or both. (2) Gunfire may hit the speeding vessel and detonate its explosives before the vessel hits the tanker, but such an explosion could rip large holes in tanker and ignite the internal LNG tank.	(1) Increased probability of civilian injuries or deaths. (2) Physical damage to the harbor and potentially severe economic or market disruptions.

Table 4.1—Continued

Scenario	Circumstances	Availability of Time	Persons Being Confronted	Objective(s) of U.S. Forces	Uncertainties About Circumstances and People	Consequences/ Implications of Deadly Force
Securing an airfield	A security forces squadron defends an Air Force base airfield outside the continental United States. Two groups of suspicious individuals are observed on the perimeter of the base. Two men have crossed the perimeter and are advancing toward the airfield with suspicious packages.	Time is an urgent factor: The security forces need to quickly assess and respond (likely within minutes) to the threat posed by the approaching individuals.	(1) Two groups of suspicious individuals on the perimeter of the base. (2) The men approaching the airfield with suspicious packages.	(1) Physically deter the individuals from entering the airfield. (2) Protect the airfield and aircraft from any potential threats. (3) Disable the approaching men and the contents of their packages if the men do not comply with commands to stop.	(1) The intentions of the individuals at the perimeter. (2) The content of the packages that the approaching men hold.	(1) Increased probability of civilian injuries or deaths. (2) A negative reaction from the population, which could damage diplomatic relations.

Table 4.1—Continued

Scenario	Circumstances	Availability of Time	Persons Being Confronted	Objective(s) of U.S. Forces	Uncertainties About Circumstances and People	Consequences/ Implications of Deadly Force
Preventing Mass Atrocities						
Confronting child soldiers	U.S. forces assisting a host nation in a COIN campaign are confronting child soldiers among the combatant forces. U.S. forces enter a village that was recently attacked by insurgents and are confronted by a group of approximately 50 armed insurgents, including child soldiers.	U.S. forces face a timecritical situation that requires a nearimmediate response to a confrontation likely to be both unpredictable and lethal.	(1) Insurgent leaders and indoctrinated child soldiers. (2) Terrorized villagers who are also related to the child soldiers.	(1) Persuade the child soldiers to turn in their weapons and either return home or turn themselves over to the authorities. (2) Minimize casualties (specifically to the child soldiers) once U.S. forces are confronted with hostile fire.	(1) The extent of the indoctrination of the child soldiers and their allegiance to the insurgency. (2) If gunfire broke out, it would be impossible to engage the adult insurgents without shooting at the children. (3) The response of the local villagers. If the children are injured or killed in the exchange of fire, this could ignite anger and hostility in villagers (who are likely related to the child soldiers).	(1) Increased probability of civilian injuries or deaths. (2) Possible alienation of the population; potential diplomatic and political fallout caused by the death of children.

Table 4.1—Continued

Scenario	Circumstances	Availability of Time	Persons Being Confronted	Objective(s) of U.S. Forces	Uncertainties About Circumstances and People	Consequences/ Implications of Deadly Force
Stopping citizen-on-citizen violence	As part of a UN peace-enforcement mission, U.S. forces must respond to the following scenario: A large shipment of arms, which is prohibited under an existing cease-fire, is discovered. The host country's presidential plane is shot down and a coup ensues. Presidential guards are opposed to U.S. involvement, but the prime minister must be protected. Escalating violence spreading to smaller urban areas and villages beyond the capital must be stopped.	Although time-sensitive decisions are not required, U.S. forces must act expeditiously and coordinate their operations over hours or days.	(1) Presidential guards found with the cache of weapons and guards belonging to the prime minister's security detail. (2) Coup leaders—soldiers, national police, militia members—meeting with villagers to instruct them on the conduct of violence and the distribution of arms. (3) Citizens who perpetrate violence.	(1) Detain the shipment of arms and prevent its distribution. (2) Protect the life and whereabouts of the prime minister. (3) Detain coup leaders at meetings and extract information on planned violence and the distribution of arms. (4) Use appropriate force to stop the escalation of violence; minimize civilian casualties.	(1) The presidential guards' loyalties and intentions about the arms cache and protecting the prime minister. (2) The intentions of citizens participating at meetings held by coup leaders: Are the citizens willing or unwilling participants in the planned violence? (3) The level of organization of citizen violence and killings in the smaller urban areas and villages. (4) The intentions of citizens participating in the actual killings: Are the citizens willing or unwilling participants? (5) U.S. forces' inability to differentiate attacker from victim in the confusion and movement of the ensuing violence.	(1) Increased probability of civilian injuries or deaths. (2) Undermined role and leverage of U.S. forces and the UN in the existing peace-enforcement mission. (3) Potential diplomatic and political setbacks to the establishment of a lasting peace agreement.

Table 4.1—Continued

Conducting Peacekeeping

Scenario	Circumstances	Availability of Time	Persons Being Confronted	Objective(s) of U.S. Forces	Uncertainties About Circumstances and People	Consequences/ Implications of Deadly Force
Operating within restrictive RoE	As part of a UN peacekeeping mission, U.S. forces conduct nightly curfew patrols. The RoE are extremely restrictive to minimize the possibility of violent incidents.	The current situation is not very time-sensitive, but U.S. forces are expected to be prepared for rapid reaction in the event of suspicious activities or an attack.	(1) Civilians who are violating the curfew and may be hostile to U.S. forces.	(1) Enforce curfew regulations. (2) Conduct operations within the RoE and minimize the frequency and scale of violent incidents and any resulting civilian casualties.	(1) The intentions of civilians and the threat they pose.	(1) Increased probability of civilian injuries or deaths. (2) Negative impact on the peacekeeping mission; potential diplomatic and political setbacks to the establishment of a lasting peace agreement.

Table 4.1—Continued

Scenario	Circumstances	Availability of Time	Persons Being Confronted	Objective(s) of U.S. Forces	Uncertainties About Circumstances and People	Consequences/ Implications of Deadly Force
Patrolling an urban neighborhood in support of local police	U.S. forces assist local security forces in securing an urban area against insurgents. During a joint U.S.-local patrol, forces come under fire from insurgents inside buildings who are likely using women as shields as part of their tactical assault to manipulate the U.S. response.	U.S. forces are under fire and will be required to make split-second decisions during this assault.	(1) Insurgents conducting an assault. (2) Women who are possibly being coerced into participating. (3) Civilians in the vicinity of the insurgent attack.	(1) End the insurgent assault and capture insurgents to gain information on future attacks. (2) Minimize the risk of civilian casualties, specifically to women who are possibly being used as human shields and to other innocent civilians in the area.	(1) The number and identity of insurgents who are hidden in buildings and among the local population. (2) Whether women are being coerced or are willing participants. Either way, U.S. and local forces will be reluctant to use force against them.	(1) Increased probability of civilian injuries or deaths. (2) Possible alienation of the population if innocent civilians are killed in the crossfire, particularly if it is confirmed that women were indeed being used as unwilling human shields.

Table 4.1—Continued

Protecting U.S. Officials and Property

Scenario	Circumstances	Availability of Time	Persons Being Confronted	Objective(s) of U.S. Forces	Uncertainties About Circumstances and People	Consequences/ Implications of Deadly Force
Resisting large-scale mob attack	Local police and National Guard units are mobilized to quell a domestic civil unrest situation in a large U.S. city as hundreds of protestors attempt to disrupt a WTO meeting.	National Guard units likely have minutes to hours to control the protestors, but as hostility escalates and leads to violence, the National Guard's response will become increasingly time-critical.	(1) Hundreds of protestors blocking key intersections, chaining themselves together to form lines, and attacking the police with baseball bats and similar club-like weapons. (2) Less-hostile demonstrators holding rallies and teach-ins.	(1) With minimal civilian injuries, control the crowd of protestors and the potential for escalating violence. (2) Prevent the crowd from defacing and destroying city property and disrupting the commercial activity of the city. (3) Protect WTO ministerial conference delegates and the location of WTO meetings.	(1) The volatility of the situation and how quickly violence may escalate. (2) The ability of authorities to control an extremely hostile and violent crowd with minimal use of lethal force.	(1) Increased probability of civilian injuries or deaths. (2) Potential severe and negative backlash against authorities; damaged relations and mistrust of police and National Guard units by civilians.

Table 4.1—Continued

Scenario	Circumstances	Availability of Time	Persons Being Confronted	Objective(s) of U.S. Forces	Uncertainties About Circumstances and People	Consequences/ Implications of Deadly Force
Ensuring convoy security	U.S. forces assisting a host nation in a COIN campaign regularly conduct road convoys between their forward operating base and a nearby airfield. Risks to the convoy are considerable and include insurgent IED attacks, small-arms fire, and ambush using civilian vehicles. A civilian vehicle pulls out of a side street and begins to drive along the convoy.	This scenario is time-critical: U.S. forces have only seconds or minutes to determine the threat to the convoy and respond accordingly.	(1) Possible insurgents in the civilian vehicle traveling close to the convoy. (2) Civilians in the vicinity of the convoy attack.	(1) Disable the vehicle and determine the threat it poses. (2) Use force that minimizes the injury or death of any innocent vehicle occupants and other civilians in the vicinity of the potential IED detonation.	(1) Whether occupants in the vehicle are combatants or innocent civilians. (2) The driver's reason for traveling so close to the convoy and the potential threat that the vehicle and its occupants pose.	(1) Increased probability of civilian injuries or deaths. (2) Possible alienation of the population.

Table 4.1—Continued

Scenario	Circumstances	Availability of Time	Persons Being Confronted	Objective(s) of U.S. Forces	Uncertainties About Circumstances and People	Consequences/ Implications of Deadly Force
Supporting Domestic Civil Authorities						
Preventing large-scale looting following natural disaster	National Guard and federal military units deploy to a large U.S. city to maintain law and order and assist in disaster relief in the aftermath of a hurricane. The situation is grave: There are breakdowns in the provision of services, transportation, and communications, and numerous incidents of large-scale looting and crime have been reported. RoE to restore law and order are extremely restrictive in this highly volatile situation.	National Guard and federal military units likely have minutes to hours to restore law and order, but if levels of looting and crime escalate, decisions will become increasingly time-critical.	(1) Armed civilians involved in looting and robbery. (2) Angry, anxious, hysterical, and hostile citizens who are desperate to obtain basic necessities, locate family members, and evacuate from filthy and insecure emergency centers.	(1) Restore law and order and security with minimal civilian casualties to ensure effective disaster relief and search-and-rescue efforts. (2) Use minimal force to calm innocent civilians who have become hysterical and possibly hostile.	(1) Risk of hitting innocent civilians in any exchange of gunfire between looters and soldiers. (2) Response from local civilians, which could range from mass panic to anger to hostility. The hurricane has created a desperate situation and left citizens with urgent needs.	(1) Increased probability of civilian injuries or deaths. (2) Any images or reports of soldiers killing or injuring civilians could create a severe negative backlash against federal and state authorities and the military.

Table 4.1—Continued

Scenario	Circumstances	Availability of Time	Persons Being Confronted	Objective(s) of U.S. Forces	Uncertainties About Circumstances and People	Consequences/ Implications of Deadly Force
Preventing unauthorized border crossings	National Guard units are mobilized to supplement the U.S. Border Patrol along the U.S.-Mexico border to apprehend and arrest persons suspected of drug trafficking and illegal border crossings.	U.S. officials are compelled to act quickly in a critical window of minutes to hours to apprehend armed or unarmed suspects.	(1) Possible unarmed civilians in vehicles or on foot attempting to unlawfully cross the border. (2) Possible armed civilians involved in drug trafficking.	(1) With minimal force, conduct border patrols to apprehend and arrest civilians suspected of drug trafficking and illegal border crossings. (2) Discourage drug trafficking and illegal immigration along the border by reinforcing border patrols.	(1) Ability to disable or stop fleeing vehicles or suspects short of using lethal force. (2) Ability to determine whether occupants of any fleeing vehicle are unarmed illegal immigrants or armed drug traffickers.	(1) Increased probability of civilian injuries or deaths. (2) Potential for diplomatic and political setbacks.

The General Analytic Framework

Because the need for a continuum of force could vary as security conditions, U.S. strategy, and U.S. military missions change, it is useful to sketch a general analytic framework before identifying requirements. The first step is to construct four taxonomies that cover

- the persons engaged, in terms of (1) types and (2) numbers
- the effects desired, in terms of (3) types and (4) intensity.

(1) Types of Persons Engaged

It is not enough to distinguish enemy fighters from noncombatants. Instead, the following categories of persons, each of which poses different problems, should be identified:

- *Dangerous.* These enemy fighters, terrorists, killers, "martyrs," and other persons are intent on doing harm, have high fear and pain thresholds, and may not be easy to stop. They may also be highly vigilant and intent on escaping, if warned, to fight another day.
- *Difficult.* These antagonistic, potentially dangerous persons are intent on frustrating U.S. objectives but are not highly trained, directed, or disciplined. They may be easily provoked.
- *Ambivalent.* These uncooperative persons have not chosen sides and may be difficult or friendly. They may want to avoid involvement and harm, and they are unlikely to be tolerant of the use of force.
- *Friendly.* These persons are inclined to cooperate or at least comply. They would like to see U.S. forces succeed and they may look to U.S. forces for security. They may be willing to experience some discomfort (but not serious pain or injury) if necessary to neutralize persons who endanger them.

It is relatively easy to conceive of nonlethal options that could be useful when the persons engaged are known to be all of one type (e.g., dangerous) or another (e.g., ambivalent), or of two types that are closely related (e.g., dangerous and difficult). The problem is that recognizing and distinguishing among such groups of people—even the dangerous from the friendly—is difficult in circumstances such as those captured in many of the scenarios previously discussed. This is especially true when dangerous persons are trying to look like ordinary friendly ones. In some cases, a collection of people may be distinguishable but mixed in type; in others, it may be mixed and indistinguishable. A group may be homogenous but its disposition and intentions unknown. Because the desired effect may vary depending on the persons engaged, the implication is, as suggested earlier, that time, information, and skilled decision-making will be required to identify, distinguish, and separate persons and administer differentiated effects.

(2) Numbers of Persons Engaged

It matters a great deal whether U.S. forces are dealing with an individual, a small group, or a large group. It is obviously easier, though not necessarily easy, to determine the type of a single individual. It is also easier to select and then administer a desired effect. At the other extreme, a large group typically consists of multiple types of persons. Figuring out which types are represented and which persons fall into those types is clearly a much more difficult challenge. Whatever the danger of administering an unwanted effect on an individual, that danger increases sharply as the number of individuals—and thus the likelihood of mixed types and degree of uncertainty—increases. Given the significance of the number of persons engaged, having nonlethal capabilities that allow forces to reduce large groups to small groups, small groups to a few persons, and a few persons to single individuals would help significantly in reducing uncertainty and risk while increasing effectiveness.

(3) Types of Effects Desired

It is not sufficient to express requirements in terms of, say, temporary incapacitation without exploring the desired effects on the behavior of the persons engaged. We have identified five types of effect, presented from least to most severe:

- halt: prevent from approaching or leaving
- disperse: cause to flee

- compel: cause to take a specific action
- control: cause to follow any order
- disable: render unable to function.

(4) Intensity of Effects Desired

Each of these effects can vary in intensity, as shown in Figure 4.1. As noted earlier, intensity may be as crucial at the low end (e.g., in dealing with unfriendly but not difficult people) as at the high end (e.g., in dealing with difficult but not dangerous people). Intensity may be an especially sensitive issue when the people engaged are of unknown or mixed different types.

A Matrix of Requirements

From these four taxonomies, we can construct a matrix showing that different types of persons (the "who") must be engaged in different ways (the "what") with different intensities (the "how much") (see Figure 4.2). For our purposes, it is reasonable to subsume the variable of the numbers of persons engaged (the "how many") under "who." The matrix can reveal, for example, whether small numbers of dangerous persons are mingled with large numbers of friendly persons.

As shown by the cells marked with an *X*, some combinations are of greater interest than others. Friendly persons may not have to be compelled or controlled, much less disabled. It would be good not to have to disable difficult persons if options to halt, disperse, compel, and control them exist. There is no limit to the types or intensity of effects suitable for use on dangerous persons, nor is there much advantage in

Figure 4.1
Effect Intensity Can Vary

	Low		High
Halt:	slow	⟶	stop
Disperse:	gradual	⟶	rapid
Compel:	influence	⟶	compliance
Control:	submission	⟶	cooperation
Disable	temporary	⟶	permanent

RAND MG848-4.1

having low or moderate effects if the persons engaged are known to be dangerous.

It is clear from this matrix that the type and intensity of the effects desired vary with the type (and implied numbers) of persons engaged. For instance, high-intensity measures that disable may be indicated against dangerous persons but not against difficult or ambivalent persons, much less friendly ones. Yet, the identity, composition, attitude, and intentions of the persons may be unknown, very complex, or both. It follows that the general analytic framework must include the variables of this matrix and confront the difficulty of knowing where in the matrix a given situation falls.

Confronting Uncertainty

One of the most critical problems in contemplating requirements for a continuum of military force is uncertainty about the prospective human subjects—a problem of knowledge, not of effects. This problem is especially pronounced at the outset of a given situation, since time, information, and subsequent events may clarify who is who.

Figure 4.2
Matrix of Requirements

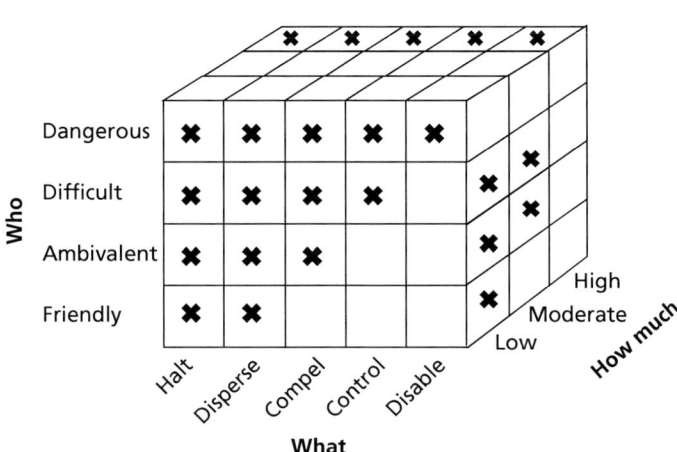

A given type and intensity of force will have reasonably predict-able effects on persons of known identity and intentions. But when a person's type is *unknown*, uncertainty about what effects are desired and what effects are likely arises. Effects intended for dangerous per-sons tend to be different from those intended for friendly or harmless persons. Moreover, the way in which persons respond to force of a given type and intensity may differ according to their intentions. For instance, enemy fighters are more likely than curious onlookers to tol-erate discomfort. A central finding is that unless these uncertainties are addressed as an integral aspect of a continuum of force, the continuum may prove ineffective and risky—not a big improvement over binary choice.

This "who problem" is illustrated in Figure 4.3. Assume for the sake of illustration that there are only two categories of persons (rather than four): dangerous and friendly. In many situations, the composition of the group of persons engaged will be neither homogenous nor clear. Therefore, whatever type or intensity of force is administered could prove either altogether inappropriate or inappropriate for some of the persons engaged. If the force is geared toward dangerous persons, for example, it could harm friendly ones (who may constitute some or all of the engaged group). In this case, the risk of harming friendly persons could outweigh any potential benefit. If force is avoided due to such uncertainty and risk, dangerous enemy fighters could gain an advan-tage (e.g., they could strike or flee). The longer dangerous and friendly persons are exposed to the same effects, the more likely it becomes that either the former will gain an advantage or the latter will be harmed. Therefore, solving the who problem quickly is crucial.

The "what problem" stems from the fact that actual effects are not entirely predictable. Although the laws of physics are predictable, effects may vary with physical conditions. They may also vary with how persons engaged react physically, physiologically, and psychologi-cally as individuals and as groups. A group of curious people is less likely to resist the use of force than a group of angry ones. Thus, the predictability of effects depends on learning who is being engaged. As Figure 4.4 illustrates, reactions may be

**Figure 4.3
The Who Problem**

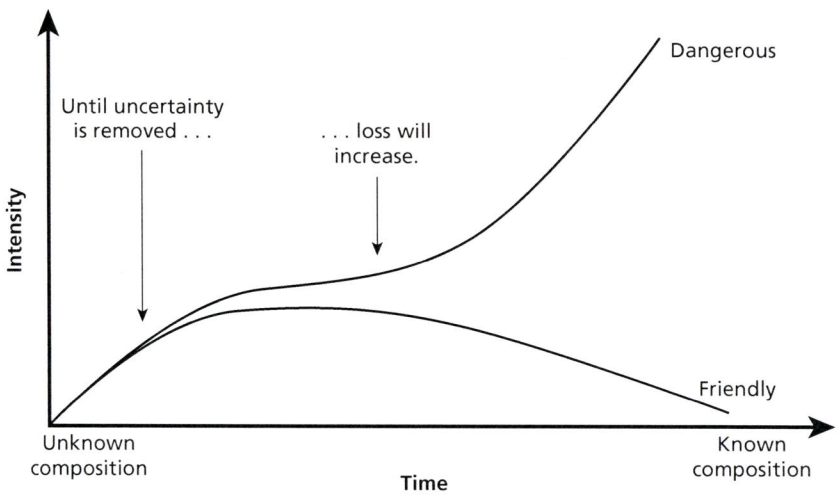

**Figure 4.4
The What Problem**

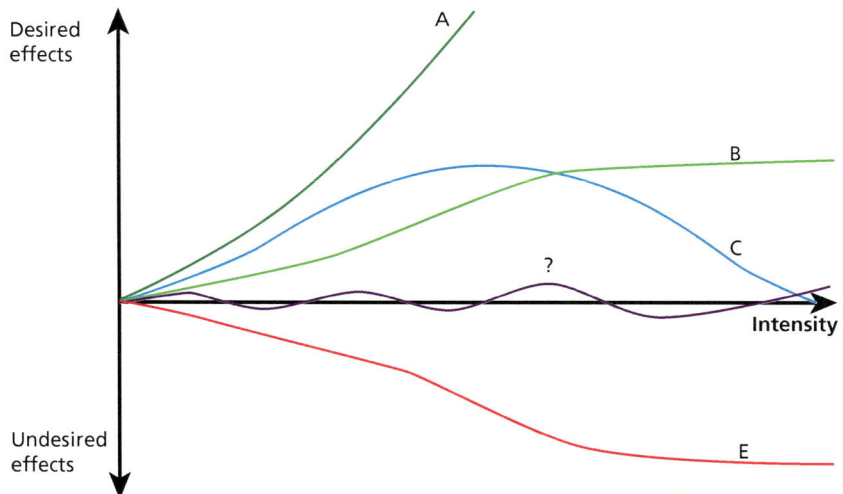

NOTE: Timing, strength, direction, duration, and differentiation of effects cannot be predicted with precision because of physiological, psychological, and situational variables.

A. quick and significant
B. slow and attenuated
C. temporary (if those engaged believe the force to be temporary)
D. negligible
E. the opposite of what is intended (e.g., if an attempt to disperse a crowd angers it instead)
F. markedly mixed (especially if the persons engaged are of mixed types).

Given that there may be a gap between intended or predicted effects and actual effects, it would be helpful to be able to sense the actual effects as early as possible. If there are doubts about the actual consequences of an administered effect, a predisposition toward using particular effect types and intensities may take over. Military forces, conditioned to destroy enemies, might be inclined toward more-severe or higher-intensity effects. Police, conditioned to minimize violence, might be inclined toward less-severe or lower-intensity effects. In many, but certainly not all, of the scenarios we considered, the risks of alienating a population (combined with the who problem) indicate the need for a police-like bias. In any case, acquiring information to address the what problem may be as critical as solving the who problem, especially if both problems are present. Of course, the need to acquire information may be in tension with the urgency to act. Therefore, a police-like bias presupposes an ability to gain information quickly.

As shown in Figure 4.5, the aim is to maximize information as a function of time. There are two ways to do this: (1) increase the collection of information and (2) gain—in effect, slow—time. Obviously, gaining time to gain information and then exploiting information to gain more time is the best way to increase effectiveness and reduce risk.

Initial Effects

These considerations lead us to an important observation: Requirements for a military continuum of force must include what we call *initial effects*. Initial effects are critical because the threshold for the use

Figure 4.5
Using Time (T) and Information (I) to Resolve the Who and What Problems

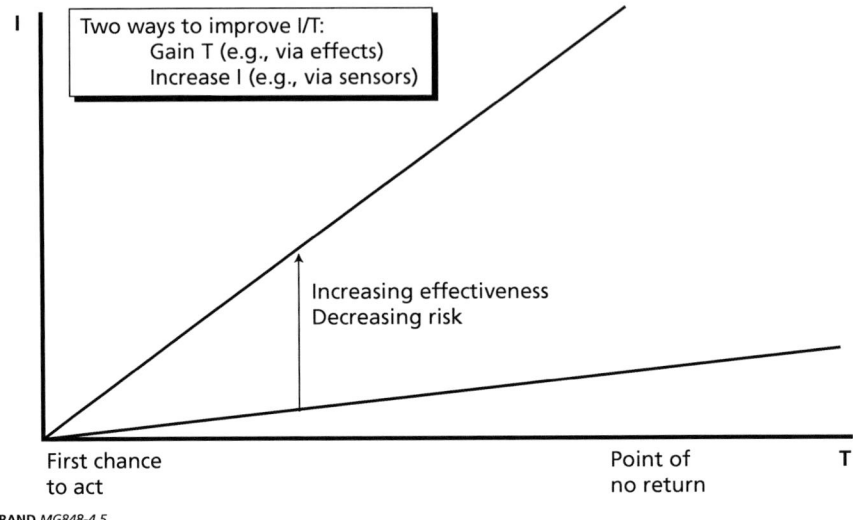

RAND *MG848-4.5*

of nonlethal force, unlike the threshold for the use of lethal force, is "very low and non-specific."[1] This being the case, it is as important to consider effects at the lowest levels of violence as it is to consider effects just below lethality. Indeed, mastering the challenge of administering initial effects through better capabilities, skills, and decision-making can provide an ability to "respond decisively . . . while maintaining personal safety in situations where deadly force is not required or is contraindicated."[2] Conversely, the inability to cause successful initial effects, whether due to indecision or inadequate capabilities and skills, can turn situational uncertainty into paralysis, mistakes, or opportunities for the enemy.

The purposes of initial effects are to

- gain time in order to get better information
- gain the initiative

---

[1]   Shupe, "Nonlethal Force and Rules of Engagement," 2003, p. 43.

[2]   Shupe, "Nonlethal Force and Rules of Engagement," 2003, p. 44.

- improve control
- extract information about who is engaged
- extract information about how those engaged may respond to force.

We do not wish to imply that there will be a clear break or interval between initial and subsequent effects. We simply suggest that initial effects should be thought of differently than effects sought when uncertainties about the who and the what are resolved.

Consideration of the operating requirements of the initial effect should not start at the lethality threshold. In keeping with the principle of minimizing harm to innocent persons, the initial effect should be mild if such persons are likely present. At the same time, the effect should disadvantage enemies. At the very least, the effect should not advantage enemies by, for example, providing warning that could enable them to flee, detonate bombs, take hostages, or otherwise seize the initiative. Threading this needle—causing no harm to innocents while disadvantaging enemies—is one of the greatest challenges in developing operating concepts and capabilities for the continuum of force. Traditional nonlethal weapons do not appear to pass this test insofar as they cannot disadvantage enemy fighters without harming innocent people.

In some circumstances, the desired effect on the persons engaged may be simply to halt or to disperse them. Both are very basic effects that are reasonably predictable, feasible, quick, and observable. Because the initial effect is intended to be mild, it may be that the requirement is merely to slow (or, conversely, cause the movement of) the persons engaged. Indeed, because the initial effect is mild, its advantage may lie not in the reaction it causes but in the information it yields.

As important as, if not more important than, the physical results of the initial effect is the opportunity it affords to gain information. When asked for the secret of his success as a battlefield commander, Napoleon is said to have responded, "Engage the enemy and see what happens." In our context, the effects of administering force, from mild to lethal, should provide information. For instance, the reactions of the persons engaged may provide information: The innocent may flee, and enemies may take a stand or attack; or, the innocent may freeze and enemies may flee. The result depends on the approach taken.

Rather than relying on reactions to the initial effect, U.S. troops need sensing capabilities to help them take advantage of the opportunity to learn both who they have engaged and how those persons are behaving. The value of the initial-effect concept depends greatly on the ability to gain information quickly.

The initial effect may also provide an opportunity for signaling and communicating. There is no set formula for what to signal: It depends on what if any warning one wants to provide, keeping in mind that it may be difficult to warn innocent people but not dangerous ones. The warning may include instructions and presume that innocent people will follow them and dangerous ones will not. Communications systems that permit exchange with friendly persons could be especially useful if they help identify enemies in mixed groups or otherwise clarify the situation. Thus, sensors, signaling systems, and communications links and devices can be highly useful during the initial phase.

The concept of an initial phase is important in identifying capability requirements, but it may not be necessary or possible in practice. Urgency and danger may indicate severe measures from the outset, and adequate information may permit severe measures at any point. However, because U.S. troops require the option to deliver initial effects as described here, the associated capabilities are needed.

If the initial effect serves its purpose, subsequent effects can be more focused and severe. At that point, who the engaged persons are, whether the group is mixed, whether types are or can be distinguished (or, better yet, separated), and how persons may react will be more clear. With effective initial effects, subsequent severe measures, if indicated, can be employed with less risk and more effectiveness. Such subsequent effects may go beyond merely halting or dispersing people. For instance, they could compel engaged persons to take a particular action (e.g., get on the ground or raise their hands) or follow orders, or even incapacitate them. In sum, the dual objectives of not harming innocent people while at the same time disadvantaging enemies may be achievable.

Given the likely mildness of initial effects, it is important to signal that effects could increase in severity. This is a general argument for scalability, wherein the potential to escalate is inherent—a prospect readily communicable to the persons engaged, be they enemy fighters or innocent people.

This discussion of initial effects has underscored an earlier observation that decision-making is important for a continuum of force. This theme is discussed in the next section.

Decision-Making

Situations that require a continuum of force tend to be very dynamic. In uncertain, fluid, and sensitive circumstances, effective reasoning and decision-making are crucial. Yet, such circumstances tend to militate against such decision-making, favoring either intuition or prior scripting instead. The problem with heavy reliance on intuition is that intuition depends on experience, which may not be useful in unfamiliar situations. The problem with prior scripting, detailed guidelines, checklists, and the like is that they cannot anticipate the potentially consequential nuances of a given situation.

Making good decisions under these circumstances requires a combination of intuition, guidelines, and training; an ability to analyze new information despite urgency; and observation and adjustment. Figure 4.6 depicts the sort of decision-making that is required. It is essential to acknowledge that the time available for this process might be so severely compressed that the discrete stages actually run together.

To illustrate adaptive decision-making, assume that a small-unit commander is ordered to seize a shipment of weapons from trucks dispersed in traffic along a vital transportation artery. However, the importance and volume of the legitimate flow of traffic precludes a hard roadblock, and trucks therefore must be slowed, not stopped. Despite warning signs and announcements, truck traffic does not slow. The commander uses a method to slow but not disable the trucks. At the same time, a message is sent out to all cell phones to announce that any

Figure 4.6
Adaptive Decision-Making

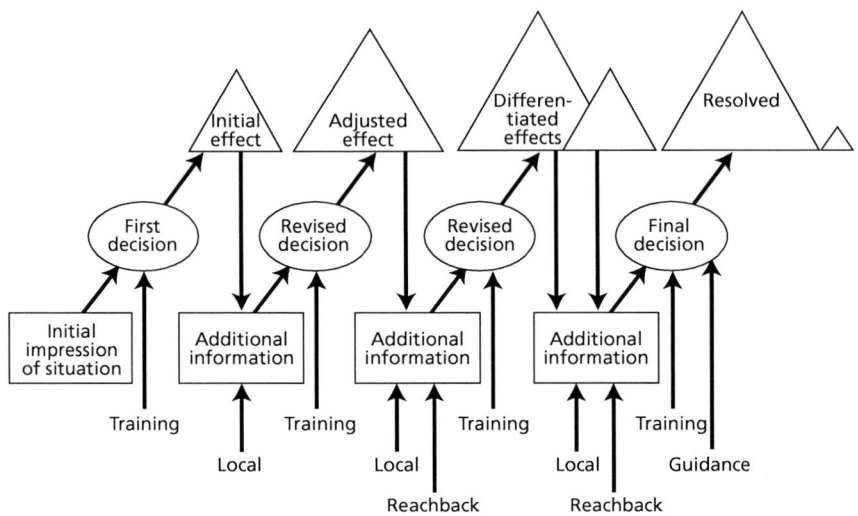

truck may be stopped and searched. Presumably, drivers who are not carrying contraband will slow down, while those who are carrying contraband will maintain speed, accelerate, or turn around. Indeed, this is the desired effect that will differentiate the enemies from the innocent. After this initial stage, more-severe effects are directed against all trucks that do not slow down. Further differentiation then allows the commander to render immobile trucks that continue or accelerate. The drivers are removed from their vehicles and, if they resist, are disabled. If they pose a mortal danger, they are killed.

Of course, any opportunity to acquire information bearing on these decisions should be fully exploited. Better yet, capabilities to acquire information ought to be designed into continuum-of-force solutions. Again, knowledge—information, communication, analysis, decision-making—is as important as physical effects.

Analyzing the Scenarios

This general framework can be used to illuminate requirements from the study's scenarios. The scenarios were selected to provide a diverse range of circumstances on which to base requirements for capabilities. Even with this diversity, six broad themes, described below, emerge.

Operations amid populations often involve small-unit engagements and a corresponding requirement for junior leaders to be able to judge, under pressure, what measures to take. The sensitivities and dangers associated with the scenarios we studied will reinforce the predilection in the U.S. military to deny small-unit leaders the authority to make decisions and instead require them to seek senior judgment. Yet, there are three reasons not to depend on higher command. First, time may be too precious to seek and await a decision from headquarters. Second, leaders on the scene may have more knowledge (including what is called *tacit knowledge*) than do their distant superiors. Third, once any level of force is used, the ensuing action-reaction cycle may provide no time for distant decision-making. For these reasons, the unit at hand needs the capability to act, including using nonlethal options. There may be no time to call and wait for backup forces with specialized capabilities. Without prejudging the appropriate level at which decisions such as these should be taken and executed in a given case, U.S. forces at the point of engagement should, as a general requirement, have the authority, information, decision skills, and capabilities to produce suitable and timely effects. Among other things, this may have an impact on the characteristics of the capabilities to be fielded (e.g., their weight, maintainability, hazardousness, ease of use).

The immediate goal of U.S. forces involved is often to gain initiative and control. In many situations, hostile forces have the initiative, if not control. This is especially the case when enemies have hostages or use human shields (willing or unwilling) whom U.S. forces do not want to harm. The commander may need to freeze or destabilize a situation to create more-favorable terms or at least to gain better control. Being able to calibrate capabilities to use force may be critical to the ability to produce effects that shift the initiative to U.S. commanders and units. Recalibration may be as important as calibration in maintaining control. In this regard, the effects of force may have to be sustained or adjusted.

Time is critical. More often than not, urgent choices must be made quickly to avoid jeopardizing mission objectives, the people or structures being protected, or U.S. troops themselves. Enemies may have more control over timing than do the U.S. forces at hand. Because decision-making is compressed, there may be a tendency to rely on intuition, which can be unreliable (especially in unfamiliar circumstances). When faced with an individual, group, or crowd whose intentions and capabilities appear menacing or threaten to become so, the commander typically needs time to decide what actions are appropriate. Time gives the commander a number of options, including communicating with the persons engaged, assessing options, collecting more information, preparing for effective action, and calling for backup forces.

Information is incomplete, ambiguous, confusing, or deceptive. The identity, motivations, and intentions of persons engaged are often unclear. Enemy fighters and innocent people may be indistinguishable from one another, mixed together in a group, or both. A group may include dangerous, difficult, ambivalent, or friendly individuals. Therefore, the problem of trying to have the desired effect can be aggravated by inadequate and possibly misleading information about the types of persons engaged. In addition, actual effects may vary from predicted effects because of uncertainty about the type and physical, physiological, and psychological response of the persons engaged, and about group and situational dynamics. Therefore, gathering information is critical to selecting the right course of action and managing what ensues. Defusing a hostile confrontation gives the decision-maker an opportunity to assess what is happening, identify dangerous elements in a group, or seek guidance or additional forces. The way in which individuals or groups react to the application of force may help discriminate between dangerous elements who have come prepared for violence and persons who are either bystanders or hostile participants with less commitment to violence. Enemy fighters may be most likely to endure nonlethal force, least likely to heed warning, and most likely to react with deadly force. Alternatively, they may be the first to flee when warned. In any case, it is essential that knowledge of a situation be improved, as opposed to degraded, with the passage of time.

Communication with persons engaged can have great value. One senior U.S. commander averred that "the best non-lethal weapon is the megaphone."[3] Our examination of the scenarios suggests that warning, persuasion, appeals, empathy, and other expressions are important but possibly insufficient. Dangerous and unfriendly persons may not heed such communications, and even ambivalent and friendly ones may be too confused, frightened, or conflicted to do so. While communication may obviate the need for force (up to and including lethal effects), it may also be needed to make the continuum of force more effective. All else being equal, force should permit, as opposed to preclude, communication.

The consequences of actions taken can be severe, complex, and far-reaching. Because of global connectivity and the worldwide media, harming innocent or ambivalent persons may have wider and more-lasting ramifications than ever before. Understanding these risks and weighing them against the consequences of failing to use sufficient force impose major cognitive challenges, especially when time and information are scarce. For example, the seizure of an HVT typically demands a prompt and clean extraction before his or her supporters can transform the situation into one much more perilous for all. A gathering crowd, hostile or not, can obstruct the extraction, especially in an urban environment; yet, firing on such a crowd can generate more animosity than even the capture of the HVT can justify. As a general rule, the type and intensity of force applied should create a high probability of achieving the task and a low probability of unwanted harm.

Conclusions

Four general conclusions stand out from this chapter. First, *U.S. troops in small units need the capability to create mild—dissuasive and disorienting but, if possible, nonviolent—initial effects in order to gain time,*

---

3   Author discussions, U.S. Special Operations Command, Tampa, Fla., 2008.

information, initiative, and control. The techniques and technologies associated with the initial effect should be able to cause a mild effect that minimizes harm and pain to innocent people while disadvantaging enemies. The effects could be as simple as slowing or causing movement.

Second, *portable capabilities are essential.* The need for small units to possess capabilities to deal promptly, even urgently, with a wide assortment of contingencies, which may or may not be foreseeable, underscores the importance of portability. Capabilities that are too cumbersome, too expensive, too scarce, or too difficult to use, and must therefore be called in, will not suffice. Again, this does not mean that such capabilities have no utility. Rather, it means that they may not be useful, timely, or available across a wide range of plausible circumstances.

Third, *scalability is essential.* Scalable effects would permit rapid escalation or de-escalation as warranted by the presence of dangerous or innocent people. They would also permit calibration, allowing for gradations between severe and mild effects. Scalability also would facilitate discrimination when both dangerous and innocent persons are present and are distinguishable or separable. Given the importance and variability of the number of persons engaged, scalability might also provide flexibility in this regard. As a practical matter, scalable-effects capabilities may obviate the need for awkward or time-consuming transitions from one set of equipment to another. Finally, scalability implies a reduced number of different systems that may be needed, thus resulting in better portability and supportability.

Fourth, *an integrated solution encompassing knowledge, skill, and actual effects is needed.* This suggests requirements for

- physical, physiological, or psychological effects
- information sensing
- signaling
- two-way communication
- training and suggestive doctrine
- educating the population
- remediation and mitigation

- cognition (i.e., intuition, reasoning, and adaptive decision-making).

Combining these conclusions with those of the preceding chapters, we can make several observations. First, U.S. policy and strategy demand a military continuum of force. Second, that continuum must provide for nonharmful or nonviolent effects. Third, a versatile, portable, scalable capability that equips small units with these effects is required. Fourth, this capability is needed sooner rather than later. Fifth, advances in information, police experience, and scalable technologies make a continuum of force possible. Finally, resolving uncertainty is as important as producing physical effects. The next chapter looks at technological options for creating this capability.

CHAPTER FIVE

Technology Options

Having established that a continuum of force, as a total system, will require (1) capabilities that produce effects, (2) information and communication, and (3) user skills, the search for options can begin with the first requirement: the means of affecting physically, physiologically, or psychologically the functioning and behavior of persons of concern.

A number of technologies that lend themselves to producing such operationally relevant effects are being pursued by the military research and development community, and by JNLWD in particular. However, we did not restrict our investigation of relevant technologies to options already being pursued; for example, we also examined work that supports law-enforcement activities. We focused on technologies that could plausibly meet operator needs by allowing users to effectively communicate, coerce, or even incapacitate people and, in some cases, vehicles. Each system examined in this chapter appears to be generally feasible, but in some cases, more work is needed to establish its efficacy, safety, and human-effects parameters. In most cases, the technology is well understood; in others, engineering challenges to incorporating the technology into a militarily useful system must be resolved.

The Active Denial System

The Active Denial System (ADS) emits a beam of directed microwave radiation that penetrates the outer layer of exposed skin down to 1/64 inch, inducing a painful sensation of intense heat. Persons exposed during testing have found the pain intolerable and have fled from the beam's cross-section. Its range is tens of meters, and the beam can be kept on for minutes at a time. The system can be focused on a particular individual, in which case its effect is near-instantaneous, or swept slowly across a group of people to ensure a dwell time of one or two seconds on each person.

The intensity of the beam can be scaled by raising or lowering the power or by spreading the beam. The effect can also be attenuated by sweeping the beam across a crowd. The faster the sweep, the shorter the dwell time and, thus, the lower the intensity. High power levels can burn skin and cause extreme discomfort, making it impossible for a person to perform a task that requires coordinated motor skills. However, the water in human tissue limits the absorption layer available to the beam and thus makes it difficult to kill someone promptly.

The current ADS system has poor portability. It can be transported only by a large, multiwheeled, Stryker-like vehicle. Future versions, if successful, may be transportable on smaller vehicles. The system's size also limits its versatility. Although it can be used to disperse a small crowd or zero in with accuracy on a particular troublemaker, its bulk limits its utility in constricted urban environments. Moreover, its high cost per unit—probably several million dollars in procurement funds—will limit the number of systems that can be deployed.

ADS technology is well developed, although more research is needed to bring down the system's size, weight, and cost. A prototype has been tested under relatively realistic operational conditions, including the irradiation of soldiers.

ADS testing has revealed that the system causes no permanent damage and that pain ceases rapidly. Potential countermeasures include wet clothing or a shield (e.g., a large piece of wet cardboard), but someone wearing obviously inappropriate clothing can be assumed to be looking for trouble.

Persons affected by an ADS squirm in obvious pain until they have danced out of the beam. This, combined with the fact that most populations are unfamiliar with the system, poses the danger of insurgents showing videos of U.S. or allied forces using a "death ray" on local citizens. Use of this system would have to be preceded by an awareness campaign. If time were limited, the use of a broadcast cell-phone message to the people in the region could mitigate the appearance of cruelty in using an ADS.

Tasers

A taser typically uses a replaceable cartridge containing a compressed-nitrogen propulsion system to deploy two small probes that are attached to the taser by insulated wires. The taser transmits electrical impulses through these wires into the remote target from point-blank range to approximately 10 meters, which is well short of the 100-meter range that is required for nonlethal systems to be broadly useful in military operations.[1] The electrical pulses affect the sensory and motor functions of the target's peripheral nervous system, causing pain and temporarily disabling the target's coordinated motor functions. A taser is intended to disable the victim for a brief time after application. It can be hand-held or fired from a weapon. In law-enforcement use, someone who is tased is then arrested (i.e., restrained by more-permanent means).

Tasers are not inherently scalable, although charging the devices beyond their normal parameters can increase the intensity of their effect all the way up to lethal force. But current models do not permit this increase to be accomplished on the spot: Doing so requires advanced intent.

Tasers are highly portable, affordable, and routinely carried by police. They are designed for use against an individual target, so a group of people cannot be tased by a single apparatus. This limits the taser's versatility to one-on-one or very few–on–very few encounters. Current versions of tasers can penetrate clothing (up to approximately 1 inch). A troublemaker could come prepared with heavy protective clothing as a countermeasure, although this would single him or her out of the crowd for a possible escalation of force.

---

[1] One hundred meters is considered "rock-throwing distance"—the range a weapon must have to afford the user protection against rock-throwers.

Broadening the applicability of tasers depends on increasing the weapon's firing range and accuracy, which could be difficult. Moreover, since there have been occasional deaths correlated with excessive use of the taser, careful calibration of the effect to be administered is needed. Finally, a person who has been tased typically drops to the ground and exhibits great, though fleeting, pain. According to one expert, "It's like being hit with a sledgehammer."[2]

Dazzlers

A dazzler is a laser, typically operated at low power, that scatters light (e.g., off the windshield of a vehicle) and makes it difficult for the target to see objects ahead. This frustrates the target's attempt to steer a vehicle, virtually eliminating his or her ability to navigate around barriers, aim and fire a weapon, or carry out a complex operation in general.

The technology allows for a measure of scalability. The laser can be operated at very low strength (e.g., to get someone's attention) all the way up to very high strength (e.g., to risk blinding—though not killing—a person).

Dazzlers are highly portable and can be gun mounted and carried by a soldier. The weapon is not particularly versatile, however. Its thin beam is usually aimed at individuals, although multiple vehicle occupants can be dazzled if the light is shined on the windshield. Thus, a dazzler can be useful in disrupting the advance of a suspicious vehicle toward a checkpoint, but it is not practical to dazzle a crowd out in the open.

The technology is well understood, and its feasibility has been proven in the field. It is inexpensive enough to be widely distributed among infantry soldiers who man checkpoints. Dazzlers now in use have an effective range of hundreds of meters, and future development should extend that range. The weapon's effects are continuous, because as long as the operator can hold the dazzler beam on a target, the person

---

2   Colonel John B. Alexander (ret.), quoted in Geoff S. Fein, "Non-Lethal Weapons Find Their Niche in Urban Combat," *National Defense*, March 2004, p. 15.

illuminated will be affected. The dazzler is hard to defeat as long as its beam is trained on the target. Its light can be attenuated by smoke and fog, but in those conditions, the target's visibility will also be degraded. The phenomenon of dazzling is not particularly exotic, and the person dazzled does not show signs of unusual pain.[3]

Femto-Second Lasers

The femto-second laser is a very-high-cycle laser that produces extremely short pulse widths and very-high-energy pulses that can be used to produce plasmas and impulse on illuminated targets without causing extensive thermal damage.[4] Highly preliminary research suggests that the biological effects from the interaction between the laser and the target might be similar to those produced by flash-bang devices (i.e., flashes, noise, and a heat-pulse that disorient and discomfort) when plasmas are generated at a moderate distance from a target. In close proximity to the target, the laser might be able to produce taser-like effects by inducing currents in the target's nervous system. The laser technology itself is relatively well understood, although it is far from being developed into a tactical weapon system, and its biological effects are not as well characterized. As a directed-energy weapon, the femto-second laser would be limited to producing line-of-sight effects. However, attenuation through the atmosphere would not be great, so its range could conceivably reach several hundred meters, well beyond the range of more-conventional devices, such as tasers.

---

[3] Note that a laser of sufficient strength could cause eye damage and even blindness, which could violate international law. (The Fourth Protocol to the Inhumane Weapons Convention, entered into force in 1998, bans the use of laser weapons specifically designed to cause permanent blindness. See Massimo Annati and Ezio Bonsignore, "Non-Lethal Weapons: Possibilities, Programmes, Perspectives and Problems," *Military Technology,* No. 27, July 2003, p. 49.) This is unlikely to be an issue with low-power lasers. Because their pulse is so short, femto-second lasers are not necessarily injurious. Nevertheless, the potential to cause blindness must be a consideration in the development and employment of high-power lasers.

[4] A femto-second laser is a laser with a pulse width of 10^{-15}–10^{-13} seconds.

A single femto-second laser device might be capable of producing a range of useful effects if its intensity is manipulated. In theory, the intensity of the beam would be scalable within a limited range through adjustments to device's the power or the diffusion of the beam through some medium. (This has yet to be demonstrated, however.) As such, the laser used at lower power could be used to warn or signal a crowd at a distance. Closer up, or at higher power, it could cause thermal damage or burns, allowing for more-severe effects.

The portability of a femto-second laser system is difficult to predict because the system is still in the early research-and-development phase. At worst, a light-vehicle mount is a realistic goal. Over time, a man-portable version could be possible but would probably require a sacrifice in terms of the number of engagements possible due to power constraints. If the research goal of a light vehicle–mounted system (or better yet, a man-portable system) is achieved, the system has the potential for broad applicability. Its only obvious limitation could be cost, and it is too early to tell whether this will indeed be an issue. The cost of some laser systems has decreased dramatically over time, while other systems have remained very expensive.

Even if the preliminary research pans out and demonstrates that the system can produce the desired biological effects, a femto-second nonlethal laser weapon system is still a long way from being packaged into a tactical prototype system that could be tested the way the ADS has been tested. Basic scientific research and development is still required to determine if this technology can be packaged in a way that allows its fielding in a tactical unit. Cooling and volume requirements in particular need additional work.

Like the ADS system, the femto-second laser has the flavor of the exotic. An awareness campaign would have to precede its application and possibly be supplemented by on-the-spot notification of any targeted crowd via broadcast cell-phone messages.

Sound Arrays

A sound array can be used to get attention and to communicate at long distances of up to hundreds of meters. As such, it can usefully communicate a warning message beyond the dangerous range of a rock-throwing crowd, the effective range of a handgun, or the damage range of an explosives-laden speedboat.

Sound can be scaled up in intensity to produce discomfort or even a moderately disabling effect. With enough decibels of power and the beam broadened a bit to affect a crowd, people's discomfort would likely cause them to disperse, although troublemakers might resist the distress. At the very least, the discomfort would make it difficult for such troublemakers to operate normally, let alone execute a complex operation. Moreover, having discriminated themselves from the rest of the crowd, they could be singled out for more-compelling measures.

A sound array can be transported on a light vehicle, but it is unclear whether further development could make it man-portable.

Intense sound has an inherent versatility as it can be used against a single actor or a crowd and can be used to communicate as well as cause discomfort. It can "broadcast" continuously. Its feasibility has been demonstrated in the field, and its cost has come down to several thousand dollars per unit.

Intense sound can be highly unpleasant, but it is familiar. The target does not appear to be enduring cruel pain. When the sound is shut off, the persons affected return to normal and do not typically suffer lasting injury. At very high decibel levels, deafness can result.

Kinetics

The purpose of a kinetic "blunt" (e.g., a rubber bullet or a rifle-propelled beanbag) is to cause trauma and pain—without lethal effect or serious prospects of permanent damage—with the intent to make victims disperse. The typical range of a blunt is tens of meters.

Kinetic rounds are not easily scalable. One potentially useful technology employs sensors to determine how far ahead the nearest target lies and reduces the firing power to ensure that the projectile does not cause permanent injury. (This is particularly important if someone cuts into the firing range unexpectedly.) Kinetic rounds are easily carried by individual soldiers because they are typically used in the same gun that fires hard bullets. Rubber bullets allow a nearly continuous rate of fire, whereas blunts are intermittent. Both can be countered by body armor, although those who are not affected by the rounds have clearly come prepared and thus have marked themselves for closer scrutiny or an escalation of force.

Kinetic systems have a measure of versatility in that they can be applied against an individual or a group of individuals. Although an individual rubber bullet or blunt is meant for a single person, shooters are usually indifferent about which person is actually hit when trying to disperse a crowd. Unintended effects depend on where the individual is hit and how robust his or her health is. At short range, rubber bullets can cause permanent scars; hits to the head or groin can be deadly.[5]

The technology of kinetic rounds is well established. The rounds have been used widely by, for example, law-enforcement units in the United States and other countries and UN peacekeeping units. In addition, affordability is not a problem. Although the sight of armed and uniformed troops firing on crowds with kinetic rounds looks a lot like the sight of armed and uniformed troops firing on crowds with live ammunition, the victims of kinetic rounds rarely bleed and almost always walk or run away.

Tear Gas

Tear gas is the most common and well-known chemical used by law-enforcement officers to disperse an unruly crowd. Coupled with the tearing effect that the gas produces, the malodorant fumes make it very difficult for people to stay in the affected area. Tear gas is rarely fatal, although people have been injured and very infrequently killed by flying canisters.

---

[5] Annati and Bonsignore, "Non-Lethal Weapons: Possibilities, Programmes, Perspectives and Problems," 2003, p. 48.

Tear gas and related malodorants are not scalable. One can disperse more or less tear gas into an area, but its general effects will not vary. Such weapons do have the advantage of being easily portable, however. A limited supply can be carried by a single troop while a backup supply is transported by a light vehicle.

This class of agents is not particularly versatile. The agents' effects cause a crowd to disperse, but such weapons cannot be used to control an individual troublemaker or compel him or her to follow instructions. The agents typically have a range of tens of meters, which is the distance a canister can accurately be fired. A patrol of soldiers' rate of fire is limited by the need to reload a new canister after every firing. Such weapons are, however, very affordable; thus, they are found in the inventory of most sizable urban police forces.

The effects of tear gas are very uneven, depending as they do on the direction and strength of the wind. A strong wind can disperse the gas and, in the worst cases, blow it back at the troops. Gas masks can counter the weapons' effects, although people wearing gas masks have singled themselves out as troublemakers by coming prepared for confrontation. A number of household remedies, most notably a handkerchief soaked in lemon juice, can mitigate the weapons' effects.

Tear gas and related agents are noxious in their effects but are not unfamiliar. Their use could fuel anger but would not be fertile material for rumors that the forces have used exotic death agents. There is disagreement about whether the use of tear gas and related agents violates the Chemical Weapons Convention, which says that riot-control agents may not be used as a "method of warfare" but does not define this phrase. The United States has long maintained that tear gas should be allowed for certain defensive purposes, such as when civilians are being used as a screen in an attack. It is obvious that even this interpretation, however, could limit the utility of tear gas in military operations.[6]

---

[6] This issue is discussed in Annati and Bonsignore, "Non-Lethal Weapons: Possibilities, Programmes, Perspectives and Problems," 2003, p. 46, which also notes that a variety of other chemicals and irritant agents are banned under the Chemical Weapons Convention.

Anti-Electronics

Anti-electronics are used to stop vehicles by interfering with their electronics and making engines shut off, all without destroying such electronics permanently. The power of such weapons can be scaled to increase the range of effectiveness up to several hundred meters, but the outcome sought (e.g., stalling an engine) does not change.

Today's anti-electronics units are not portable. They are both relatively heavy and expensive, although the latter problem can be mitigated if long-range stopping is sacrificed. With enough power, the units can affect multiple vehicles simultaneously, although in such cases, destructive force may be more appropriate.

This class of systems lacks versatility. It can be very useful in specific scenarios, most notably the need to stop a vehicle that rushes a checkpoint and ignores warnings to halt. However, this is not the most demanding sort of scenario that occurs in many types of military operations. Furthermore, the weapons' ability to achieve effects depends on the kind of engine being targeted. Cars built more than 25 years ago tend to contain fewer electronic systems and are therefore less vulnerable to anti-electronic beams.

Although this technology application is not familiar to the general population, the effects appear relatively benign because the damage occurs to vehicles rather than people.

Flash-Bangs

The purpose of a flash-bang is to disorient victims and leave them temporarily unable to hear or see very well. These effects can break up a threat by preventing coordinated action. While disoriented or frozen in place, potential enemies can be subject to more-forceful action. Flash-bangs are not scalable, however: An individual flash-bang has a single setting.

These weapons are highly portable and can be easily carried by individual soldiers. They are relatively versatile because they can play a role in disrupting and disorienting a small crowd or, less ideally, an individual. Moreover, since they are used in small quantities, they are highly affordable. Typically, flash-bangs are used at short ranges of approximately ten meters.

The application of flash-bangs requires an element of surprise. Properly prepared, targets can avoid their effects by wearing earplugs and looking away from the flash. In general, being hit by a flash-bang is not painful, and the effects are usually temporary.

Flash-bangs can be jarring, but their effects are not mysterious. The potential stigma associated with their use is limited by the fact that flash-bangs are a surprise and their effects have dissipated before anyone realizes what has happened.

General Observations

There appears to be no silver bullet that is scalable from mild to severe as well as versatile, portable, and feasible. Described above are a number of technologies that operate along at least a part of the warning-discomfort/disorienting-disabling-lethal continuum. Some demonstrate considerable military utility in a variety of situations, but no one technology is useful across the entire range of contingencies represented by our scenarios. Thus, the greatest payoff in the near future will hinge on developing an integrated suite that combines two or more capabilities whose applications are complementary and which in combination cover a broad spectrum of scenarios.

With one partial exception—femto-second lasers, discussed below—the current technological trajectory for scalable-effects weapons is one of continuous improvement in specific parameters. For most of these technologies, improvements in power-to-weight ratios are critical. They can mean the difference between a system that can be transported by a heavy vehicle only and one that can be mounted on a light vehicle, or the difference between a vehicle-mounted system and a man-portable system.

Another important and appropriate thrust in current research across the range of scalable-effects weapons is improving our understanding of their effects on humans.[7] Such weapons can be deemed unusable if they "overachieve" (e.g., by harming innocent people) or "underachieve" (e.g., by failing to disable or disorient a dangerous enemy in time). A good example of overachieving, for instance, is provided by reports that people have died after having been tased and confined.[8] Even if such results are rare, caused by attendant conditions, or exaggerated, they raise questions about the predictability of a particular alternative, which could degrade its value. Such uncertainties, especially when associated with new systems, can compound the risks of enemy propaganda and political backlash, which are addressed in the next chapter.

Perhaps there are other promising technologies that neither DoD nor this study has uncovered. Indeed, it is important that the search continue not only for applicable known technologies but also for less-well-understood phenomena that could help meet the need for a continuum of force. Whether in assessing current technologies or in searching for others, the general framework, specific requirements, and assessment criteria employed in this book can be useful and, we hope, will be regarded as an enduring contribution to the field.

---

[7]   Of course, many scalable-effects systems also work against things. For example, microwaves can shut down electronic devices, and lasers can put tiny pits in windshields.

[8]   See also Mark Kroll and Patrick Tchou, "How Tasers Work," *IEEE Spectrum*, December 2007, pp. 24–31.

CHAPTER SIX

A Promising Approach

Assessing the Alternatives

The preceding chapter identified options to enable small military units operating amid populations to defuse threats, including unexpected ones, and to carry out their missions without having to kill or harm noncombatants. Although the list of alternatives identified is not exhaustive, it is indicative, and thus permits preliminary conclusions about which technologies are most promising.

The alternatives can be assessed according to criteria based on strategic and policy requirements (from Chapter Two) and operating requirements (from Chapter Four). As previously described, we drew operating requirements from numerous representative scenarios across the following types of important noncombat missions:

- COIN
- humanitarian intervention
- peace operations
- protecting U.S. officials or property
- supporting civil authorities in quelling domestic disturbance.

These assessments are shown in Figure 6.1. Green indicates a positive assessment, red a negative assessment, and yellow a mixed or uncertain assessment. In these assessments, we place special importance on four criteria, shown in the figure in bold. We consider these four criteria essential in the sense that a general solution to the continuum of force must in large part be

Figure 6.1
Assessment of Alternatives

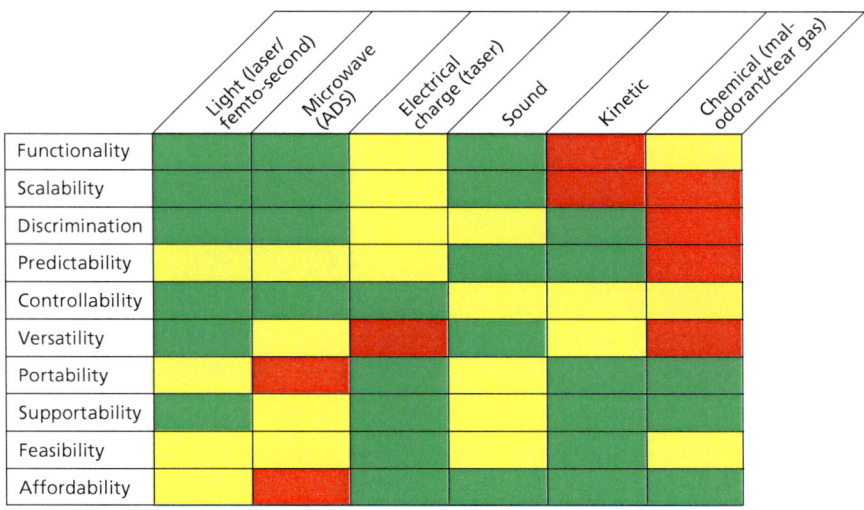

RAND *MG848-6.1*

- *scalable* from mild, nonharmful effects (to induce caution) to severe effects (to cause incapacitation) to lethal effects
- *portable* by a small unit, ideally on foot but at least in light vehicles
- *versatile* across a wide range of conditions (including being useful at rock-throwing distance)
- *technologically realistic* (already extant or at least based on practicable science).

This stress on scalability, portability, versatility, and feasibility does not detract from the importance of the other criteria. For example, inadequate controllability—for example, the potential to injure or panic accidentally large numbers of people that need only be warned—could rule out an option whose risk of misuse or abuse is disproportionately large given its advantages. Similarly, lack of predictability in the effects upon and reactions of the persons engaged by a capability may lead to adverse results or inhibit the use of that capability.[1] If a capability costs too much, it will not be bought in sufficient numbers to be carried regularly by the standard small unit.

With this in mind, one can weight each capability's scores according to the importance of the criteria, perhaps manipulating those weights based on different theories of which criteria are most important. In any case, alternatives with poor scores in the scalability, portability, versatility, and feasibility criteria may be disqualified as candidates for the general solution sought by this study. (As previously noted, this does not mean such technologies are not worth pursuing for specific purposes.)

Note that no single option stands out above all others or completely satisfies all of the four key criteria. Because each of these criteria is necessary but none is sufficient, it follows that there is no obvious single solution to the general problem. Several options could be useful under, but *only* under, limited conditions:

- microwave (e.g., ADS), when portability or assured availability is unimportant
- anti-electronics, when vehicles must be disabled
- flash-bang munitions, when confined spaces must be entered
- tasers, when a few individuals at very close range must be immobilized
- rubber bullets or tear gas, when forces are confronted by concentrations of moderately hostile persons.

These capabilities might be deployed with and used by forces when such conditions can be anticipated, or emplaced where they are likely to be needed. However, *they do not offer a general solution to a wide range of problems that small ground-force units on the move might encounter across a range of important missions*, including COIN, stability and peace operations, operations to quell civil disorder, and other noncombat circumstances.

---

[1] Some experts and senior officers believe that predictable, reliable, repeatable non-lethal weapons effects are not merely important but are in fact essential to allowing field commanders to have enough confidence to order their use (Bedard, "Nonlethal Capabilities: Realizing the Opportunities," 2002, p. 5).

More specifically, a typical small unit patrolling a potentially hostile and crowded neighborhood without knowing whether and which threats could arise cannot routinely be accompanied by a large ADS vehicle. Either the threat would have to be anticipated or the operation would have to be delayed so that ADS could be deployed. Anti-electronics capabilities are of no value when potentially threatening persons are not in vehicles or do not otherwise rely on electronics. Flash-bang munitions are limited in range and could be viewed as threatening by nonhostile persons who happen to be present. Tasers are very limited in range and thus in versatility: They are useful only against small numbers of individuals and may not offer adequate scalability. Tear gas may antagonize otherwise nonhostile persons who are mixed in with hostile ones. Rubber bullets are incapable of mild effects, can cause serious injury and scars, and are not scalable.[2]

Although most of these alternatives have some utility in some circumstances, this research is not intended to match such specialized capabilities with the limited, specific applications for which they are suitable. Rather, our aim is to identify capabilities that are of such wide utility across important missions and possible conditions that general-purpose units can be trained and equipped with them.

A Promising Approach

The two options of greatest general interest involve directed energy in the form of *sound* or *light*. The sound option involves tightly-focused sound waves of scalable decibels that are conducive to conveying voice or other sounds. The light option involves tightly-focused light beams (up to and including lasers) of scalable power and intensity.[3]

---

[2] Many of the problems with rubber bullets stem from improper use or accidental discharges that lead to rounds being fired too close to the target or at a vulnerable point (such as the head).

[3] Because a laser is not simply a bright beam of light but a focused beam of coherent single-wavelength energy, scalability of light from bright light to laser is problematic, if not infeasible.

Intuitively, such capabilities are broadly attractive. Directed energy is inherently scalable and versatile. Both sound and light are reasonably predictable as phenomena and in their effects. Unlike some other options, neither depends on physical projectiles or contact (which can be counterproductive if used against nonhostile persons). At the same time, both act upon the senses—hearing and sight—on which enemy fighters critically depend. They can confuse, disorient, incapacitate, or dissuade without necessarily causing harm. Assuming that they ultimately prove feasible (see below), the two options display important qualities:

- smoothly calibrated scalability from mild (nonharmful) to severe (permanently harmful) effects
- sufficient range (i.e., well beyond rock-throwing distance) for most situations in which nonlethal means may be indicated
- useful against large groups, small groups, and individuals
- virtually sure to have an effect because of limits on human sensory tolerance for directed energy
- capable of sowing confusion, if desired
- useful in separating or at least discriminating between determined or dangerous persons and others, thus reducing the number of targets and making individual targeting possible
- conducive to observing, gauging, and adjusting to actual effects
- useful in getting immediate attention.

It is not difficult to imagine countermeasures against directed sound and directed light, which could include seeking physical shelter, covering the ears, and covering the eyes. But such countermeasures would likely impair the actions of, and limit the danger from, targeted persons. Both sound and light offer a way to generate *initial effects* (as prescribed in Chapter Four) to gain time, gain information, sort targets, and escalate as dangerous and innocent persons are separated physically or by their responses. Make no mistake, directed sound and light can cause harm—actual or perceived, temporary or lasting. Either one can anger innocent persons or be exploited for propaganda by adversaries. Yet, the ability to sharpen both the focus and the effects as enemy fighters are identified or isolated offers a way to take more-forceful action without hurting noncombatants.

At the same time, neither light nor sound will suffice as a general solution. Sound can provide a continuum of effects in a narrow or wide swath against a number of persons at extended range, but it cannot be easily used with discrimination against dangerous individuals in the midst of innocent ones. Furthermore, sound might not immobilize individuals if they have shielded themselves.[4] Light, including lasers, can provide a continuum of effects at a reasonable range. However, it may not be adequate in responding to a large number of potentially threatening persons, in producing wide-angle effects, or (except for lasers) in bright sunshine.

While neither option may be sufficient, the two are complementary in sensory effects, in range, in angle width, in daytime and night-time use, and in cases when line of sight is blocked. Moreover, they could be used simultaneously for greater sensory effect. It follows that the most promising general solution may be a suite of directed sound and light (including laser) capabilities that is enabled by advanced information and decision-making abilities. While neither sound nor light is adequate to fill the void below lethal force, each is scalable, and both are complementary.

To test this approach, we reexamined a number of scenarios with the assumption that such a suite would be available to the U.S. units in question. Although the suite did not always prove to be the best solution or entirely adequate, the combination of directed sound and directed light has broad utility. In general, the scenarios suggest that

- Sound could be used at long ranges, wide angles, and relatively low decibel levels in order to alert and warn persons of concern.

---

[4] "Directional high-intensity acoustics . . . can cause ear damage at high power and close range" (Neil Davison, *The Contemporary Development of "Non-Lethal" Weapons*, University of Bradford, Department of Peace Studies, Bradford Non-Lethal Weapons Research Project, UK, May 2007, p. 21).

It could also be intensified to disperse crowds or separate non-combatants from combatants.[5]

- The effects of intense sound could be augmented and compounded by intense light.
- If smaller numbers of hostile individuals persist despite piercing sound and bright light, they could be targeted with lasers. If some harmless individuals remain among the dangerous ones, lasers may cause them to desist while inflicting no lasting harm.
- Sound or light (including lasers) could be further intensified and used selectively against dangerous individuals as such targets are identified. If these individuals seek shelter, sound may still be effective. If they find a way to tolerate sound, lasers may still be effective. If they seek protection from the energy directed at them, they may become less threatening.
- Having gained time and determined which persons are truly threatening, troops could use lethal force with greatly reduced risk of killing innocent people, thus reducing some risk of causing the deleterious consequences that such casualties can have on the larger mission. Lasers are potentially deadly at extremely high levels of power, but the unit would probably have a full array of lethal weapons from which to choose.

Although neither light nor sound has yet been shown to be man-portable, both could likely be carried by small units that possess light, fast vehicles. Because such vehicles are in any case likely to be used in most of the missions of interest here, there may be no special or additional requirement for acquiring them. It is possible that a single vehicle could carry both capabilities. In addition, there are several ways to improve the portability of the options:

- Research and development could be devoted to reducing the size and weight of the power required for sound and light systems.

---

[5] One reason directed sound is not adequate to satisfy the continuum requirement is that it is effective in significantly altering hostile behavior only when intensified to levels that can also permanently damage hearing (National Research Council of the National Academies, *An Assessment of Non-Lethal Weapons Science and Technology*, 2003, p. 31).

- Small units that might encounter circumstances in which the capabilities may be needed could be accompanied by such vehicles.
- Quick-response tactics could be developed.

An important technical question is whether directed sound and directed light can each be adequately scalable from mild to disabling effects from a single, easily portable source or device. Meeting this requirement should not be a problem with sound. Scalability from bright light to lasers from a single source or device could be a more serious technical obstacle.

Adding an Information Component: Cell Phones and Video

Earlier in this book, we highlighted the importance of combining information capabilities with effects capabilities. The reason for this is that the sorts of problems likely to be encountered by U.S. forces involve high degrees of uncertainty and ambiguity that need to be resolved before intense effects are attempted. Moreover, performance depends on resolving uncertainties quickly—more quickly than can be accomplished using effects capabilities alone. One interesting possibility is the use of cell phones to communicate with the people at hand.

Related RAND research suggests that cellular telephony holds significant promise for improving the performance of U.S. forces in COIN and other missions of interest.[6] Three things make this possible:

- Cell-phone infrastructure, devices, and use are increasingly widespread and could be expanded.
- Assured access to cell-phone switches can permit identification of cell phones with users, geolocation, and easy two-way and broadcast communication.
- Most actors—insurgents, bystanders, indigenous forces, etc.—will be using cell phones.

---

6   Libicki et al., *Byting Back*, 2008.

The specific application offered here is the ability to call all cell phones within a given geographic area. Using previously established links to the authorities who control switching, a local commander could request that all cell phones in an area of concern be called instantaneously. This would permit, at a minimum, transmission of simple text or audio messages that issue, for example, warnings or instructions. Even if only a fraction of the individuals present received the message, others present could, and surely would, be told its contents. The utility of including this capability in the scalable suite is borne out in a number of the study's scenarios.

It is possible that cell-phone identification and geolocation could provide a local commander with useful information about the persons present in an area of concern. Two-way communications would allow local commanders to receive valuable and timely information about the composition of a crowd and its intentions as well as about the effects of any weapons used. In considering directed sound and lasers, however, we are concerned only with the ability to use cell phones to send messages to all users within a given area as a complement to directed sound, light, and lasers.

The suite of capabilities prescribed here could be further enhanced by another available technology: cameras that transmit live and recorded video. Again, this is a technology that has advanced substantially and has been applied widely in recent years, mainly in the interest of maintaining security. Cameras can be carried by individuals or mounted on vehicles, on weapons, and in fixed locations that are either frequently used or critical. The potential advantages of capturing video during COIN and other operations amid populations include

- observing, sharing, and archiving what has occurred for purposes of learning lessons about, for example, crowd behavior, the actual versus predicted effects of nonlethal systems under various conditions, the effects of nonlethal systems on different types of individuals, and escalation (or de-escalation) sequencing
- collecting evidence of immediate physical effects with which to counter unfounded rumors and propaganda. In this regard, video cameras can inhibit unwarranted or risky behavior among U.S.

troops or, alternatively, encourage warranted behavior by increasing confidence that decisions will be justifiable in the face of later questioning. Video could be especially important when systems that might produce lasting injury, such as high-powered lasers or microwave, are used.

- collecting intelligence on, among other things, the dangers posed by certain persons in an area or the whereabouts of persons known to be dangerous.[7]

With additional development, cameras could provide real-time local observations on which to base tactical further moves (e.g., cell-phone messaging or scaling of effects). Video systems could also enable remote viewers of an operation to provide information, advice, or (if need be) orders to a local unit commander, especially if his or her observation or cognition is obstructed or distracted.[8] Creating communication pathways to permit this should not be an insurmountable problem if modern cellular telecommunications infrastructure is in place.[9] At the same time, achieving adequate communication capabilities may entail size, weight, and power implications—at worst, it could be like asking a unit to enter confrontations equipped with the equivalent of a television crew.

In its fullest form, then, the suite of continuum-of-force capabilities envisioned here consists of directed sound, directed light, lasers, cell-phone communication, and video observation. As a whole, this suite is remarkable in its nonkinetic character: It affects the senses and perceptions of persons engaged rather than their physical condition.

---

[7] Information obtained from cameras could be combined with information about the owners of cell phones present in the area to help determine the identity of individuals.

[8] The risk here, of course, is that higher command able to observe local developments will not resist the temptation to micromanage.

[9] Use of shorter-range radio-frequency links to vehicle-mounted local switches would be possible, but might require a more complicated approach than is desirable for a typical small unit. Creating such communication pathways may be more than just moderately challenging, however. Cell-phone video is relatively low quality and thus may be insufficient to make the kind of close distinctions that are required to support useful second-guessing by someone not on the scene.

This does not mean that kinetic nonlethal or lethal capabilities have no place in the range of options available to U.S. forces operating amid populations. Yet, the logic of alternatives to physical violence leads to a host of emerging but largely proven technologies which, if used creatively and together, offer U.S. forces ways to control situations and gain advantages over enemies without harming persons who ought not be harmed.

Political Realities, Reactions, and Risks

As already noted, force-continuum capabilities and their effects could be misrepresented by anti-U.S. propagandists or antipathetic media organizations to depict the United States as being involved in sinister new forms of warfare against innocent people (e.g., Muslim noncombatants). The possibility of nonviolent or nonharmful effects will go a long way toward neutralizing this danger. Still, the potential for distortion and unwarranted alarm should not be underestimated. For instance, one can imagine the use of lasers against civilians being reported and denounced as causing everything from blindness to sterilization to brain damage.

The key to depriving such reports of substance is to limit *actual* pain and injury. This is consistent with the standard, suggested early in this book, that such persons should not be hurt, harmed, or antagonized any more than U.S. citizens in similar situations would be. Thus, a question worth asking is whether the U.S. government would authorize use of a particular capability in the midst of U.S. populations. Yet, even that high standard does not preclude political damage abroad that is disproportionate to the actual effects of such capabilities.

An important consideration in this regard is whether a population has experience and familiarity with a particular capability and its effects. For example, while rubber bullets are painful, they are familiar and generally known not to cause lasting harm. In contrast, an unfamiliar effect from what may seem a mysterious device could cause

great consternation, abundant rumors, and lasting suspicions that subsequent ailments are the result of that device.[10]

It is possible to construct a hierarchy of the actual, psychological,
and political effects of different options both on the people directly
affected by the weapons and on those exposed to reports, rumors, and
propaganda. The variables are level of pain, persistence of injury, and
degree of strangeness. For a given level of effectiveness, capabilities that
cause neither pain nor injury, have familiar effects, and are not conducive to propaganda are better than those that fail to meet one or more
of these criteria.

Table 6.1 illustrates how such perception, political, and propaganda risks can fit into the assessment of alternative capabilities. Capability A appears to be better than both capabilities B and C *until* its
strangeness to the persons affected and its vulnerability to adverse propaganda are considered. Although capability B may cause temporary
pain, it is preferable to capability A because it is less likely to produce immediate panic, subsequent rumors, or effective propaganda.
Capability C scores highest in terms of avoiding political fallout, but it
is least effective. Because we have identified no single alternative that
scores high in every respect, the aim must be to design and develop an
integrated continuum-of-force suite of capabilities that does so.

Table 6.1
Factoring Perceptions and Propaganda into Assessments

Likely Result	Capability A	Capability B	Capability C
Is effective	High	High	Low
Avoids pain	High	Low	High
Avoids injury	High	High	High
Produces familiar effects	Low	High	High
Is propaganda-proof	Low	High	High

[10] These findings were in part the result of getting the reactions of experts on Arab and
Muslim populations to aspects of a continuum of force (various speakers, Roundtable Discussion on Continuum of Force, RAND Corporation, Arlington, Va., September 11, 2008).

Exotic devices that cause sharp pain and lingering (real or imagined) abnormal conditions are likely to alienate ordinary people, intensify the anger of unfriendly people, and be used effectively by enemy propagandists to turn a larger population against the perpetrators (e.g., U.S. forces). At the same time, capabilities that cause little pain and no harm but are themselves strange or are thought to cause strange effects could generate negative repercussions among both those exposed and the wider population.

Compare, for example, tear gas with ADS. Neither is likely to cause lasting harm. Tear gas is highly unpleasant but familiar. ADS—also known, tellingly, as the "pain ray"—may cause sharp, momentary discomfort and is unfamiliar.[11] Although both could give rise to subsequent claims of lasting harm, ADS is more vulnerable to such claims because it is unfamiliar. In addition, its use could enable propagandists to claim that U.S. forces are using experimental weapons with grave and not entirely understood effects, including, for example, sterilization, genetic mutation, and cancer.

Applying this reasoning to the general sound-light-laser solution described above, we find that

- In and of themselves, bright light and loud sound are familiar, and fabricating stories about bizarre and enduring effects would be accordingly hard.
- However, high-energy lasers would be unfamiliar, and rumors about their hideous effects could abound.

The potential for alternative force-continuum capabilities to be misunderstood or misconstrued or otherwise to lead to unfortunate political consequences is summarized in Figure 6.2, where low risk is green, moderate risk is yellow, and high risk is red. In general, sound and light are less likely to cause adverse psychological and political problems than are physical weapons, chemicals, or what could be viewed as shocks or rays (such as those produced by lasers, microwave, and tasers).

---

[11] Alec Wilkinson, "Non-Lethal Force," *The New Yorker*, June 2, 2008, p. 26.

Figure 6.2
Political and Psychological Risks

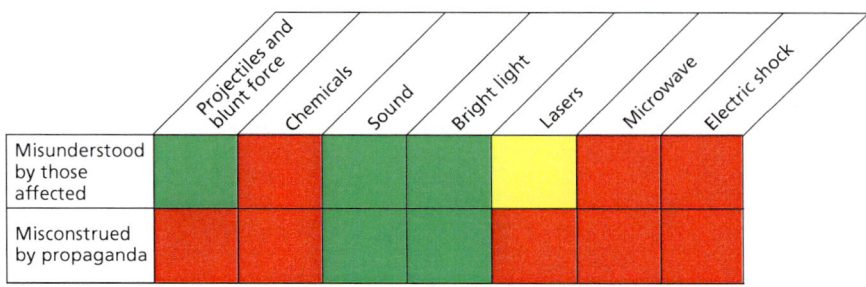

RAND *MG848-6.2*

The more unfamiliar the systems used and effects caused, the more important it is to raise awareness of U.S. purposes and capabilities. It is crucial that any message accurately communicates what U.S. forces are in fact trying to do: deprive killers the benefit of hiding within the population but at the same time minimizing harm to the population. In other words, U.S. forces take seriously their responsibilities to provide safety and avoid violence. This message should be communicated before continuum-of-force capabilities are used, and it could be reinforced by use of the cell-phone feature described earlier.

At the same time, use of cell-phone messaging to alert citizens (as well as enemy fighters, of course) of imminent actions or steps they should take also raises important psychological and political questions. Although citizens may appreciate being warned to avoid an area, to disperse, or not to be concerned because they will not be hurt, they may at the same time react adversely to the perception that U.S. forces or their own government is able to send them messages and, by implication, access their phones (and conversations). The way around this is to give them the choice of whether to avail themselves of such a warning service. Although some might decline the option, those who accept will not resent the messages but rather be reassured by them. Participants in the warning system, however small in number, could pass alerts on to nonsubscribers in a crowd. In time, more people would likely sign up for this service. Although the right to buy and possess a cell phone could be made contingent on subscribing to the service, such a policy would be unnecessarily coercive.

Likewise, many people will not appreciate increasing levels of video-camera surveillance; as readers would attest, the key to acceptance lies in being convinced of the security benefits. On the whole, there is no substitute for communicating early, persistently, consistently, and accurately the rationale behind and facts of all aspects of a continuum of force to populations that may be affected. The unifying theme of such communication must be that U.S. forces take with utmost seriousness their responsibility to safeguard the people of those countries where they operate, just as they would in the United States itself. The combination of this message and, of course, behavior consistent with it would go a long way to ensuring that a continuum of force will help U.S. forces not simply avoid mistakes but actually succeed.

Assessment

This discussion of political risks reinforces a central theme of this study: A continuum of force is as much about the awareness, reasoning, and behavior of the people using and experiencing such force as it is about hardware. Directed-energy and cell-phone capabilities are of special interest for all the reasons previously outlined, but they alone cannot satisfy the need for a continuum of force. It is vitally important to be able to sense and comprehend a fluid situation, know how to respond when the allegiance and intentions of the persons engaged are unclear, understand and anticipate behavior, communicate, and be sensitive to the wider repercussions of actions. Accordingly, as the remainder of this book addresses questions of implementation, much of its two final chapters examines how U.S. troops and decision-makers must be conditioned for a continuum of force.

With this caveat, we conclude that a continuum of force that meets our requirements is possible but complex. It entails multiple physical, informational, and cognitive capabilities and will require a good deal of attention to developing and integrating the components.

CHAPTER SEVEN

Operation, Preparation, and Organization

Notional Concept of Operation

Having found that a suite consisting of sound, light, lasers, cell-phone communication, and video observation could provide a continuum of force, we turn next to the important matter of developing a concept of operation (CONOP) to maximize the utility of such a suite in the field. As previously stated, this combination of capabilities has the potential to reduce reliance on unnecessary levels of force against the populace while still accomplishing the job at hand through a flexible sequence of operations that may include warning, sorting, dissuading, disorienting, impeding, and (if necessary) incapacitating groups who are hostile, nonhostile, or some of both. Such sequencing allows for both escalation and de-escalation as indicated by a unit's objectives and the fluid conditions it faces and senses.

Using the combination of technologies advocated here could afford a much richer set of options for ranges of military operations, such as effective low-impact crowd control at extended ranges (hundreds of yards or more) and nonlethal "overwatch" in situations in which the use of lethal force by an adversary is a possibility. However, the main advantage of this type of system comes from its ability to help security forces better control dynamic situations while avoiding counterproductive use of the most violent effects, which can undermine political and security goals.

Imagine, for example, a U.S. unit with the twin missions of establishing a control point and apprehending potentially hostile individuals as opportunities arise. The unit confronts a group of individuals, some of whom are not hostile to the United States, some of whom are fence-sitters who might be agitated to active resistance, and some of whom form a core of dedicated individuals hostile to the U.S. mission. The unit is unsure, however, of this group's composition, and it knows even less about the allegiances of individual group members. As the unit approaches, expecting to be confronted, it sets up an initial position that takes advantage of a long line of sight from elevated locations that will allow it to employ its suite of scalable-effect and associated communications capabilities.

The unit is trained, equipped, and ready to employ the previously described combination of cell-phone, sound, light, and laser systems together in a manner that lets the troops achieve their objective with no more violence than is needed and without unduly exposing themselves to attack. Generally speaking, these capabilities are going to be used for sorting the crowd—thinning out the population until only truly dedicated hostile individuals are left—and then for incapacitating the remaining individuals to allow the unit to establish physical control for identification and detention. At no point does the unit compromise its preparedness to use lethal force.

The CONOP for the use of this suite of systems is illustrated in Figure 7.1. It can be viewed as a cycle that forces restart when they advance to a new location. In the first phase of the operation (T_0), the unit (blue) uses light and sound to provide warning and thin the crowd. Because the crowd in this case is an intermingled mix of nonhostile (black), hostile (red), and fence-sitting (green) people, the main idea is to separate and remove the likely nonbelligerents in a way that avoids making the situation worse (for example, by harming nonhostile persons or converting fence-sitters into belligerents).

During this phase of the operation, forces gather information on the crowd and reduce the total number of individuals in the area through a combination of mild measures (the initial effects described earlier). Use of the cell-phone system gives forces some information about the cell phones and cell-phone users present in the immediate

Figure 7.1
Notional Concept of Operations

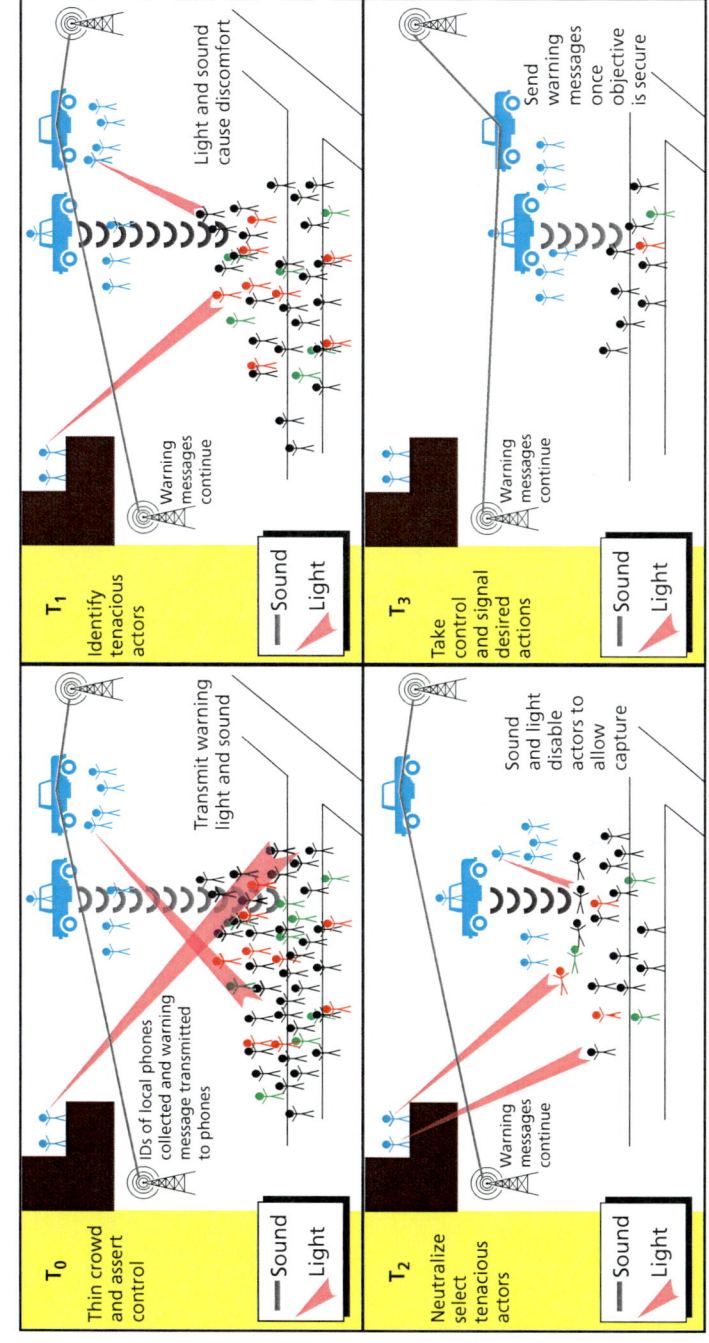

RAND MG848-7.1

area,[1] especially if nano-cells are available to help forces determine which phones are present in a small geographic area.[2] Ideally, overhead observation (e.g., via surveillance drone) provides a different perspective of the group than is obtainable from the ground and allows an accurate mapping of the area. The operations can succeed, however, without this overhead view.

After initial warnings sent via cell phone, sound, light, or some combination, the crowd has presumably been thinned somewhat. The remaining individuals, having been given ample warning, are most likely (although not certainly) hostile. The security forces have gained an idea of which cell phones are present and perhaps some information about who owns those phones. It is still not possible to determine the motivations of the crowd in general or any individuals in particular, but the circumstances allow for reasonable escalation of nonlethal force to further sort the group and permit the security forces to gain more information and pursue their objective.

During the next phase (T_1), forces begin to inflict discomfort in order to move the group in a controlled manner and further decide which persons may be of interest. Sound arrays, light, or lasers are used in a measured fashion at levels that cause discomfort and shape how the group behaves, including pushing parts in different directions and at different rates to make it travel in the desired manner and direction. The unit commander and key personnel watch for and gauge this behavior, comparing it to past experience, research, lessons learned about crowd behavior, and training (including simulations). Thus, an active-control system is applied to the group to maximize the ability of the security forces to pursue their objectives while protecting the group

---

[1] The "who" in this case might also include the actual identity of the individual associated with the unique SIM card of the phone in some cases where it has been determined through other means (see Libicki et al., *Byting Back*, 2008).

[2] This presumes a fairly sophisticated use of the cell-phone system, and the use of user and operator equipment that can provide accurate positioning information for security forces. Although noncooperative operations might be possible using some sort of signal-interception capabilities, such capabilities would be much less usable in situations in which compromising the system capabilities would not be warranted by the operational gains.

from accidents, such as stampeding or flowing into chokepoints, that could lead to fatal crushes and injuries.

Individuals who remain after this level of force is applied are evidently committed to staying, but their intentions are still not necessarily clear. For example, the scalable systems may not have caused the expected level of discomfort. Perhaps some in the group are tolerant of the discomfort. Perhaps some are determined to resist the security forces and even to attack the unit. Proceeding to much higher levels of violence based on the expectation that the group is uniformly hostile is appropriate based on the information available at this point, but it is prudent to use sufficient force that the security forces can advance and take control of individuals who present the greatest threat or interest. Again, the unit's standard package of lethal weapons can be used at any point the commander chooses. As we will soon see, only a few persons in the unit are involved in using scalable-effects capabilities, and even they can switch to lethal weapons swiftly.

During the next phase (T_2), forces use the unique capabilities afforded by a weapon such as a femto-second laser to disable selected individuals at a significant distance from the unit. The objective is to disable the dangerous individuals in the group or induce the remaining group as a whole to comply with orders. (Again, this requires knowledge about experiments in crowd modeling and about local behavior.) When feasible, such efforts would also enable capture teams from the unit to target particular individuals for retention or to gather better information. The effects produced by a femto-second laser could incapacitate multiple individuals in fairly short order, albeit selectively. At a minimum, it would add to their disorientation and degrade their ability to operate.

Selecting and affecting targets at significant distance with a disabling effect is by far the most stressing part of the operation. However, it permits forces to disable the person who was clearly coordinating the group, as well as nearby supporters who might interfere with the unit's capture teams. The key to the operation is to observe and adjust promptly to the responses of the group and its individual members. Along with the ability to interweave light, sound, and laser effects against enemy combatants in close proximity to noncombatants, the use of electric-powered scalable systems means that the force will not run out of ammunition.

In reality, these stages would not follow the set-piece format implied by Figure 7.1. Rather, they would require dynamic and nuanced coordination between force elements to create optimum effect. The interplay of security forces using a variety of devices to move crowds, sort persons within crowds, and incapacitate and capture dangerous or high-value persons—all the while being prepared to use deadly force— may be very complex. It may evoke the type of play that arises during games of constant motion, such as soccer or hockey, where the fluidity of the game makes set-piece action problematic. These operations also require forces to work well in dynamic conditions and suggests a level of expertise both in tactical command and force employment that may exceed the ability of personnel for whom these operations are an ancillary mission. In fact, the skill of operators is the most critical component of this operation once a system that allows for limited effects can be employed.[3]

The final stage of the mission (T_3) involves consolidating the security forces' position and preparing the area for future operations. The security unit uses cell-phone or sound-based communications to provide further warning or instruction. By this time, the unit has sufficient information to assess the composition and intentions of the remaining group. The unit may make further detentions or take additional measures to reassure the crowd or mitigate the unit's previous actions.

It is important to consider how and when countermeasures might be applied to disrupt U.S. forces' use of nonlethal systems. The countermeasures could be passive (e.g., protection against the direct effects of various nonlethal systems), active (i.e., measures designed to attack the nonlethal systems and render them ineffective, such as antimaterial sniper fire against scalable-effects equipment), or tactical (e.g., the use of larger numbers of people to saturate an area and overwhelm the

---

[3] Some of these effects might be achievable without lasers and other advanced systems. For example, flash-bangs could contribute to the effects needed during this sort of operation, provided they can be detonated in a pattern around people of interest. However, the laser systems that allow for more-graduated effects (ranging from simple warning to incapacitation) offer much more flexibility and potential nuance in their employment.

unit). Thus, developing a flexible and robust approach to scalable-effects systems will be as important as it has been in other types of military operations.

As it approaches these sorts of operations, the U.S. military needs to embrace the adage that the enemy has a "vote" in how the operation unfolds. The preferred U.S. approach will present a smaller number of critical nodes and, generally, smaller and more-distributed nodes; it will avoid tactics that are so overly structured and predictable that the adversary can easily plan for them and effectively react. This means that the military art and science of using scalable-effects capabilities will be at least as complicated as those pertaining to more-conventional military operations. This, in turn, underscores the need to think through carefully how U.S. forces should prepare and organize for operations that may call upon a continuum of force.

Preparation and Organization

Even in the linear terms just sketched, continuum-of-force operations, like the circumstances in which they occur, are complex, dynamic, and fraught with uncertainty. Moreover, scalable responses by U.S. forces must be nuanced, flexible, and often nonlinear as a situation unfolds. These challenges are very different and in some respects more demanding than those presented by force-on-force combat. This has important implications for preparing and organizing military units to use scalable-effects capabilities to best effect. There are also implications for the skills required for effective employment, training, organization, and command and control.

Informing all these issues is our finding that employment of a continuum of force demands *exceptionally high levels of performance and, therefore, of preparation.* Most mistakes and abuses of nonlethal weapons stem from inadequate training and supervision.[4] At the same time, as has been made clear, a continuum of force must be available for use by ordinary small units of U.S. ground forces involved in COIN, peace

---

4   Alexander, "An Overview of the Future of Non-Lethal Weapons," 2001, p. 190.

operations, humanitarian intervention, and comparable missions. This poses an obvious problem: How can the "average" regular small unit achieve such excellence in using new and nuanced capabilities in such complex operations? Fortunately, the quality of the typical Army or Marine Corps small unit, including the associated officers and non-commissioned officers (NCOs), is very good—in large part because these units have gained experience in and learned from operations in Iraq and Afghanistan. So, the starting point is high. Still, the performance demands indicated here can be met only with concerted preparation and organization.

The first issue is the skill required for effective action. For the suite of capabilities considered here, there is a relatively low barrier to proficient technical use of the systems, assuming an emphasis on usability in design. However, the question of how *best* to use the systems—including sensing, timing, and sequencing—is quite complex. These aspects will require significant training, first to master the basics and then to adapt to the locale or mission environment in which these capabilities may be employed. These requirements cannot be met merely by inserting training on the use of scalable-effects capabilities into *standard* individual or unit training.[5]

This raises the question of whom to train—a question that cannot be answered until it is determined who within a unit should be involved in the sensing, decision-making, and execution aspects of a continuum of force. There are two models worth considering. In the first, in order to stress the continuity from nonviolent force to lethal force, every member would have the ability to access and use directed sound, directed light, or both. Accordingly, every member would be fully trained in these capabilities in addition to other standard and spe-

---

[5] Training and education are important not only for operating units and individuals but at all levels. In Bedard, "Nonlethal Capabilities: Realizing the Opportunities," 2002, the author stresses the importance of such training in service and staff schools not only to improve effectiveness but also to gain acceptance for non-lethal capabilities throughout the military leadership. This view is echoed in Council on Foreign Relations, *Non-Lethal Weapons and Capabilities*, 2004, p. 6: "There is a need to integrate information and training regarding non-lethal weapons into the curricula of schools at all levels in each service; this in turn would increase the rate of integration into current force capabilities."

cialized duties. In this case, guidelines on use would have to be detailed and strict; moreover, the commander would have to exercise control over the escalatory actions of every individual in the unit. Finally, every solider would require substantial training above and beyond, yet integrated with, current requirements.

In the second model, only a few members of a unit would have the ability to access and use these capabilities. While they might also have standard duties, they would be especially well prepared in every respect to use the capabilities skillfully and wisely. They would know not only the technical aspects of the systems and their effects but also such behavioral matters as individual and crowd psychology and behavior, signs of changing danger, how to distinguish enemy fighters from nonfighters who look alike, how to determine what can go wrong, and, of course, when and how to escalate or de-escalate. Such individuals might be more-senior and more-seasoned troops (i.e., NCOs). They could be given more latitude: standards and general guidelines, not tight scripts. The commander could give orders to such a well-prepared team-within-a-unit rather than be concerned with controlling all unit members, including the least-prepared personnel. This would enable the commander to concentrate on the essential task of reasoning through and adapting to the overall situation.

The team-within-unit approach may also be advantageous inasmuch as a sizable majority of soldiers in the unit would remain more or less dedicated to the use of lethal force, should it prove necessary. The danger of inhibiting or impeding the use of lethal force has been a continuing theme in military skepticism about nonlethal weapons—a concern we certainly do not dismiss. If every member of a unit had to make the passage across the lethal-force threshold, the risk of delay or confusion would be greater than if only a few NCOs—prepared for precisely that passage—had to make it. At worst, the volume of lethal force would be reduced only by the small fraction of the unit belonging to the scalable-effects team.

Finally, as a practical matter, the need to place high-powered sound and lasers in a single vehicle fits the second model better than the first. Indeed, portability constraints may dictate reliance on a team of specialists. This does not mean that individuals throughout the unit would not use gun-mounted lasers for illumination. The point is that the main capabilities of the continuum would be under the control of a small, specialized, and experienced team.

This second, preferred approach is akin to the way the Army and Marine Corps treat other specialties within a unit.[6] Take, for example, the selection of squad or platoon "point men." These positions are critical to the success of a multitude of small-unit missions in today's conflicts. But not every soldier or marine has the ability, aptitude, and attributes to be a stellar point man, and such strengths are not easily taught. Some persons—and, thus, some soldiers and marines—are endowed with an especially acute sense of situational awareness. Their instincts, vision, and hearing are such that platoon squad leaders assign them to one of the most critical tactical roles in a unit.[7] While they have to be able to perform the same standard duties as all members of the unit, they can be entrusted with and prepared for a role that requires special qualities and abilities.

In the case of scalable effects, the situation is similar: It will take more than technical proficiency to employ these systems selectively, effectively, flexibly, and safely. It will take a high tolerance for ambiguity, an appreciation of complexity, and seasoning in dangerous situations in which reflexive use of force could lead to mistakes. Thus, the team-within-unit approach for continuum of force follows a tried-and-true pattern of specialization even in small units.

At the same time, not every NCO will be right for such an assignment. Knowledge of local conditions, experience in operating among populations, and experience in managing escalation will be critical

---

[6] This is consistent with the view ascribed to U.S. unit commanders that the best approach is to assign responsibility for nonlethal weapons to one or a few specialists. See Fein, "Non-Lethal Weapons Find Their Niche in Urban Combat," 2004, p. 3.

[7] The role of squad or platoon rear security is another role that is not awarded simply upon graduating from "rear-security school." The platoon leader and NCOs soon learn who can be trusted in this second critical role of a small tactical unit. Again, situational awareness, though hard to teach, is invaluable to a tactical unit. The skills requisite to a majority of the other positions in a small unit (other than medic and sniper, which require situational awareness in addition to classroom training) are learned during advanced infantry training and further developed through tactical workups prior to deployment.

in determining how well someone will perform in this environment. Accordingly, NCOs should be screened for all that goes with the role of using a continuum of force.

Selected and specially prepared NCOs can manage the employment of scalable-effects capabilities, but the officer in charge of the unit will be responsible for overall strategy, resource allocation, significant escalation decisions, and fusion of information and capabilities. Such a division of labor suggests that both NCOs and junior officers must be well trained in their respective roles as well as together as a team. Junior officers and key NCOs should thus be trained and educated in continuous sense-making and adaptive decision-making under conditions of uncertainty, urgency, and risk, especially amid populations.

Although scalable-effects capabilities would be treated as a specialty, every member of a small tactical unit should have some basic training in their use to ensure that every soldier has at least an elementary understanding of what the unit can do and, more importantly, to undergird unit training. Given the likelihood of continuum-of-force operations and the need for the entire unit to conduct itself properly, it is essential for unit training to encompass these capabilities and tactics. Both the Army and Marine Corps have made significant strides, such as realistic urban-warfare training, in preparing their tactical units for the broad range of mission sets that arise during today's conflicts. During predeployment workups, unit training should capitalize on information gleaned from after-action and postoperations reporting as well as integrated reporting from embedded systems (e.g., video) that capture actual operations.[8]

In the longer term, depending on advances in portability and the nature of operational requirements, every soldier might be equipped with and prepared to use scalable-effects systems. This would present greater command and control problems for unit officers and NCOs—not to mention significantly greater individual screening and training requirements—but the time may indeed come when these capabilities and the cognitive and technical ability to use them should exist at the individual level. Whether or not that is a long-term goal, treating the

---

[8]   See Libicki et al., *Byting Back*, 2008.

continuum of force as a specialty of highly qualified and more-senior persons is in any case the place to start.

A related matter is the character and content of the instructions to be given to these teams-within-teams and their commanders prior to operations. As has just been made clear, it is better to rely on a few highly trained persons than on all or most of the personnel in a given unit. A notable advantage of the former situation is that the few highly trained personnel will not require detailed or rigid instructions. We find that clear but flexible guidelines akin to those on which police departments rely are preferred. Getting CONOPs, standards, and general doctrine right is essential; scripting behavior is neither essential nor desirable.

Philosophically, the development and fielding of scalable-effects weapons should follow a "walk-before-run" process that is designed to develop (1) an understanding of the weapons and (2) the tactics and doctrine that allow for successful employment of these weapons over time. A deliberate pace is essential for introducing scalable systems into a culture whose most meaningful metric for more than 200 hundred years has been based almost wholly on lethality.

CHAPTER EIGHT

Feasibility, Integration, and Implementation

Feasibility

The two preceding chapters demonstrate that the need for a general-purpose continuum-of-force capability that is scalable from nonviolent to lethal effects could be met by a *portable suite of directed sound and light, including lasers, area-wide cell-phone messaging, video-cameras, and a well-prepared, specialized team-within-unit.* Before considering how such a suite could be integrated, it is important to be sure the technical hurdles associated with each component are known and can be overcome with reasonable effort. These components seem feasible, but a closer examination is warranted before drawing conclusions.

As already noted, directed sound may be used to hail or warn selected persons at a distance of hundreds of yards, confuse or disrupt their actions, cause them discomfort, or require them to take countermeasures (e.g., for sound suppression) that could impede their movements. Beyond that, causing serious trauma or harm would likely require very-low-frequency sound, which cannot be targeted tightly and selectively (e.g., at enemy combatants surrounded by noncombatants). Therefore, the segment of the continuum for which sound is applicable is at the nonviolent end, which is one reason why sound needs to be accompanied by other capabilities. Vehicle-portable directed-sound devices are feasible, and the same device could cover a range of frequencies and intensities. Although such systems are already being developed, they should be designed and pursued specifically as part of a scalable, portable, integrated suite usable by ordinary small units.

Light is more complex. To span the continuum of force, light might be used to illuminate areas, illuminate targets, degrade sight, sow confusion, cause discomfort, disable persons, or disable vehicles. The low end of this range could be accomplished with extremely bright light; the middle with dazzlers or high-energy ("green-light") lasers; and the high end with extremely short-pulse, high-intensity (i.e., femto-second) lasers. Of these, only the femto-second laser raises major development challenges.

As noted earlier, lasers rely on different physics than ordinary bright light. Therefore, continuous scalability across these directed-light segments may not be feasible or worth the effort required to develop such systems. One possibility is to scale down a laser by reducing its intensity through filtering, which would permit it to be scaled back up via unfiltering as a situation unfolds. Another possibility is to manipulate power to the laser emitter. However, it may not be necessary to have single-source or single-device scalability if operators can shift quickly and gracefully from one source to another. In principle, such light and laser capabilities are at least vehicle-portable, and some could be carried by individuals on foot. Standard vehicles could be adapted for sound, light, and laser capabilities.

Power is an obvious concern with any sort of directed energy. One advantage of configuring a suite is that there could be a common power supply, also vehicle-portable, for sound, light, and lasers. Other practical issues include setup time and ease of use. These and other engineering opportunities and challenges remain to be examined.

Recall that the four essential criteria for a military continuum of force are scalability, versatility, portability, and feasibility. The last of these is important because policy, strategy, and operational conditions and missions dictate that the need for the continuum already exists. All else being equal, effort should be concentrated on capabilities that involve less technological risk and development time. The technologies of light, sound, cell phones, and video are well established. The only problem that could take long to solve is that of creating practical femto-second lasers for the high end of the nonlethal segment of the continuum. Still, most of the elements of the continuum could be available in a few years.

Given the high stakes for U.S. troops and foreign civilians, continuum-of-force capabilities should be developed and fielded expeditiously but not hastily. Such systems can be the best of friends or worst of enemies to small units. If the systems are not virtually 100-percent reliable, predictable, and operationally effective, a unit will be reluctant to use them when they have other choices (e.g., lethal options). Alternatively, a few mistakes could erase the benefits of otherwise successful use, especially taking into account the pitfalls of rumor and propaganda. Scalable systems must work well every time.

Having established that each of the components appears technically feasible, the requirement for a suite obviously increases complexity and raises concerns about the feasibility of the whole. To the extent that the suite depends critically on the availability of every component, its feasibility is no greater than that of the least-feasible component (i.e., femto-second lasers). However, as long as some components are available—e.g., sound, light, "green-light" lasers—the suite could have real value, and a continuum of force could at least begin to take shape. Integrating the suite into U.S. forces should definitely not be postponed until every segment of the continuum is filled.

A more serious problem is whether the integration of these components—including the information and communications features—is feasible in the near-to-medium term, especially when taking into account the need for sophisticated operators and nuanced doctrine. Addressing this problem will require more than just technical achievement.

Integration

Again, this study found no single way to fill the void from nonviolent to lethal action. Individual approaches appear suitable for specific circumstances, but their utility across scenarios is not sufficient to qualify them as general solutions for small units operating under uncertainty against adversaries amid populations. The particular suite described here involves no critical components that are unavailable or infeasible. Still, its overall effectiveness depends on how well the components are integrated into a more or less singular functional capability. Two sorts of integration in particular are important:

- The effects must be complementary and span the void between nonviolence and lethality.
- The component technologies must be technically compatible with one another in, for example, space, weight, preparation time, logistics requirements, and operator demands.

We have already established that the weapons' effects are complementary and can fill the current void, with the possible exception of crippling or lethal effects, which may be beyond the capability of either directed sound or practical lasers. In addition, the components are similar in size, weight, and other features, and none will pose practical problems for forces bearing and using the suite that the others do not. They are portable on a small vehicle; they can be aimed and used quickly; they require significant but not enormous power; and they are readily scalable. Because sound would not interfere with light, sound and light could be used sequentially or concurrently for best effect. The fact that both sound and light involve directed energy should make it easier for military personnel to master their use and, just as importantly, avoid their misuse. A minor problem is that simultaneous audio cell-phone messaging could be impaired by the use of directed sound, and text messaging could be impaired by intense light. However, the presumption is that such messaging would occur before or after energy pulsing.

In addition to technical and operational compatibility, the components should to the extent possible be integrated for ease of transport and use. Because space and weight are at a premium for small units, common power sources, common vehicles, and even some common hardware (e.g., video monitoring or laptop computing) would all be advantageous. Common or at least similar displays and controls would permit more-seamless shifts and thus better scalability. These considerations underscore the importance of designing and engineering these capabilities *as a suite* (while still enabling standalone use).

Taking into account both these technical matters and the human and organizational factors presented in the preceding chapter, it appears possible to create a coherent general solution based on several components that fit with one another and with the structure and operating concepts of a typical small unit. Translating this possibility into reality will take institutional clarity and commitment (discussed below).

Investment and Implementation

Components of a potential continuum of force that are already in development, production, or the field—e.g., cell-phone applications, directed-sound devices, gun-mounted target-illumination lasers—should be favored in terms of attention and resources. At the same time, they should be reconceived—and if necessary reworked—to be folded into a continuum. The aspects that will require development include

- very-high-intensity[1] sound that is precise, scalable, and effective at long ranges (hundreds of yards) and can cause discomfort or incapacitation
- femto-second lasers
- software that permits selective and instantaneous cell-phone messaging to users in a particular area
- deployable links for real-time video
- improved portability of all elements of the suite, with a view toward fielding some or all with dismounted troops.

Whenever the U.S. military is faced with the need for integrated solutions, its reflex is to issue a request for proposal for a major systems-integration contract. We caution strongly against treating the continuum of force as a "project" or "program" in traditional defense-acquisition parlance. In addition to the need to keep costs down, there is a need to foster innovation, flexibility, and competition. There is also

---

[1] That is, powerful and focused.

a need to evolve and enhance the continuum with forces *in the field*. None of these desiderata—including keeping costs in check—is likely to be fulfilled by a massive systems-integration contract that takes years to compete and years more to implement. With a few exceptions, the development and production required to create a continuum of force could and should be done on a fixed-price basis with fierce competition at every stage.

Investment takes resources. In the cases of technical development, system acquisition, and system integration, the key resource is money. The funding available to the JNLWD today does not begin to reflect the importance of enabling U.S. forces to act effectively against adversaries in the midst of populations whose cooperation is needed for campaign success. We have not estimated the funding required to develop the suite described in this book. Nor do we advocate that all development take place within the JNLWD. Other labs and the Defense Advanced Research Projects Agency could engage. As for acquisition, while we have pointed out that affordability is a factor in investment strategy, we have not estimated the system, lifecycle, or total program costs.

Having said this, it is possible to make a rough judgment about the minimum level of resources that are needed and could be effectively applied in the coming years. Using JNLWD's current budget of approximately $50 million as a point of reference, we believe that budget growth of 25–50 percent per year would be easy to justify in view of the need to (1) investigate the suite of capabilities described here and (2) broaden the nature of development to include information, communications, and cognitive capabilities. A growth rate of 50 percent per year, however, would be exceptionally challenging for any government organization to achieve. The lower end of the range, 25 percent, suggests that at least $250 million in funding above the current JNLWD budget is needed over the next five years—an amount double that budgeted for the period based on the current budget. This recommendation is low compared to other independent findings about the need for additional resources in this domain.[2] Of course, the funding requirement will grow as capabilities are acquired.

In developing a continuum of force, investment in people is as important as investment in things. The primary need is for training, which also requires resources—mainly time. Unit training would not dramatically increase total unit-training time, but the individuals (i.e., NCOs) directly involved would need to be trained and retrained for months, if not years. This is a significant commitment, though one we find would be well justified.

Finally, as the U.S. military develops a continuum of force, we urge it to pursue international collaboration not only with close U.S. allies (e.g., NATO) but also with the military affairs branch of the UN's peacekeeping department and with a wide circle of like-minded countries that face similar needs. There are few if any risks associated with such collaboration, and it is in the United States' interest to foster widely the fielding and use of capabilities that can be effective against enemy fighters without harming civilians.

---

[2] The Council on Foreign Relations recommended annual spending of $300 million—a six-fold increase (Council on Foreign Relations, *Non-Lethal Weapons and Capabilities*, 2004, p. 2).

CHAPTER NINE
Conclusion and Recommendations

As is always the case with the development and deployment of a new capability, one is left facing the question, *Who is to do what now?* We recommend the use of a well-established model: the executive agent. Currently, the Marine Corps is the executive agent for nonlethal weapons. We have made clear that "nonlethal weapons" is far too narrow as both a concept and a capability. The task is to create a continuum of force from the nonviolent to the lethal in the form of a portable, versatile, scalable, and feasible general solution—an integrated suite that the typical small unit can employ. This requires

- some technical development of components, with an eye toward their compatibility and integrated use
- CONOPs and standards of use
- selection and preparation of individuals and small teams in terms of both technical and behavioral aspects.

Such a multifaceted undertaking is best done by an executive agent, and we see no compelling reason why it should not be the Marine Corps. However, if the Marine Corps, upon considering its future missions and priorities, concludes that the continuum is not of critical importance for its organization or small units, executive agent responsibilities could be shifted to either the Army or Special Operations Command (SOCOM). Tempting as it is to look to SOCOM for innovative capabilities, it must be remembered that the missions and situations in which units will require—and already require—a continuum of force are so various and frequent that ground-force units must be prepared. This argues for making the Army executive agent if the Marine Corps rejects the role.[1] Should the Army assume this responsibility, it would be essential for its Training and Doctrine Command to accept the challenge to develop the concept of a continuum of force in the broad sense defined by this book.

As an alternative, given that several services could make use of continuum-of-force capabilities, a case can be made for placing the responsibilities with Joint Forces Command. Whichever organization has this responsibility, there is a need for a concerted effort to craft CONOPs, guidelines, investment plans, training programs, and organizational measures (e.g., to enable small units). The problem of integration—not only of technical components but also of skills, doctrine, and decision-making—also argues for strong leadership and partnership with JNLWD on the part of whichever organization's leadership is committed to the continuum as a concept and its introduction into U.S. forces.

As for technical research and development, we urge a middle ground between extreme centralization and extreme decentralization. Naturally, the military executive agent will need a coherent plan with adequate resources to bring the continuum of force into being. At the same time, it is important to involve all the organizations that are most suited to dealing with particular technical facets, including both effects technology and information technology. Because technical development must be harmonized with adjustments to personnel, structures, training, and doctrine, we caution strongly against outsourcing the development of a continuum of force as a whole.

Although small units are the key to deploying and employing the suite this book describes, the continuum of force has clear implications —positive in purpose but with potential downsides—for strategy and policy, the topics with which we began our study. Therefore, the continuum should receive the attention of leaders at the highest levels. Senior officials and officers should be aware of the potential of these

---

[1] SOCOM would and should develop specific capabilities to meet those of its own specific needs that are not shared by the ground services.

capabilities, the need to shape the military culture to use them wisely, and the advantages of delegating authority.

In sum, a continuum of force for regular U.S. forces operating against adversaries amid populations is needed and feasible. Scalable and portable technologies are in train or within reach, but they do not provide a complete solution. The capability to prevail against dangerous adversaries without harming innocent people and jeopardizing entire campaigns depends critically on the skill, sensitivity, and preparation of U.S. soldiers. In turn, creating and mainstreaming this capability requires vision, initiative, commitment, and persistence on the part of those soldiers' civilian and military leaders.

Bibliography

Alberts, David S., and Richard E. Hayes, *Power to the Edge: Command . . . Control . . . in the Information Age*, Command and Control Research Program, U.S. Department of Defense, Washington, D.C., 2005. As of October 13, 2008: http://www.dodccrp.org/files/Alberts_Power.pdf

Alexander, John B., "An Overview of the Future of Non-Lethal Weapons," *Medicine, Conflict and Survival*, No. 17, Vol. 3, July 2001, pp. 180–193.

————, "Non-Lethal Weapons to Gain Relevancy in Future Conflicts," *National Defense*, March 2002, pp. 30–31. As of December 1, 2008: http://www.nationaldefensemagazine.org/ARCHIVE/2002/MARCH/Pages/Non-Lethal6828.aspx

Annati, Massimo, and Ezio Bonsignore, "Non-Lethal Weapons: Possibilities, Programmes, Perspectives and Problems," *Military Technology*, No. 27, July 2003, pp. 44–50.

Bedard, E. R., "Nonlethal Capabilities: Realizing the Opportunities," *Defense Horizons*, No. 9, National Defense University, Center for Technology and National Security Policy, Washington, D.C., March 2002.

Council on Foreign Relations, *Non-Lethal Weapons and Capabilities: Report of an Independent Task Force*, New York: Council on Foreign Relations, Inc., February 2004. As of October 13, 2008: http://www.cfr.org/content/publications/attachments/Nonlethal_TF.pdf

Davison, Neil, *The Contemporary Development of "Non-Lethal" Weapons*, University of Bradford, Department of Peace Studies, Bradford Non-Lethal Weapons Research Project, UK, May 2007. As of December 1, 2008: http://www.brad.ac.uk/acad/nlw/research_reports/docs/BNLWRP_OP3_May07.pdf

Department of Defense, Directive 3000.05, *Military Support for Stability, Security, Transition, and Reconstruction Operations*, November 28, 2005.

Fein, Geoff S., "Non-Lethal Weapons Find Their Niche in Urban Combat," *National Defense*, March 2004, pp. 14–17. As of December 1, 2008: http://www.nationaldefensemagazine.org/archive/2004/March/Pages/Non-Lethal3625.aspx

Florida Department of Law Enforcement, Criminal Justice Standards and Training Commission, *Defensive Tactics Curriculum*, "Legal and Medical Risk Summary," June 2002, pp. 1–9. As of February 3, 2009: http://www.fdle.state.fl.us/CJST/Curriculum/DTLegMedRev.pdf

Gompert, David C., *Heads We Win: The Cognitive Side of Counterinsurgency (COIN): RAND Counterinsurgency Study—Paper 1*, Santa Monica, Calif.: RAND Corporation, OP-168-OSD, 2007. As of October 13, 2008: http://www.rand.org/pubs/occasional_papers/OP168

Gompert, David C., Irving Lachow, and Justin Perkins, *Battlewise: Seeking Time-Information Superiority in Networked Warfare*, Center for Technology and National Security Policy, National Defense University Press, Washington D.C., July 2006.

Gompert, David C., John Gordon IV, Adam Grissom, David R. Frelinger, Seth G. Jones, Martin C. Libicki, Edward O'Connell, Brooke Stearns Lawson, and Robert E. Hunter, *War by Other Means—Building Complete and Balanced Capabilities for Counterinsurgency: RAND Counterinsurgency Study—Final Report*, Santa Monica, Calif.: RAND Corporation, MG-595/2-OSD, 2008. As of October 13, 2008: http://www.rand.org/pubs/monographs/MG595.2/

Ignatius, David, "Fighting Smarter in Iraq," *Washington Post*, March 17, 2006, p. A19.

Joint Non-Lethal Weapons Program, *Developing Capabilities*, 2008. As of October 13, 2008: https://www.jnlwp.com/developing_capabilities/default.asp

———, *Future Capabilities*, 2008. As of October 13, 2008: https://www.jnlwp.com/future_capabilities/default.asp

Kroll, Mark, and Patrick Tchou, "How Tasers Work," *IEEE Spectrum*, December 2007, pp. 24–31.

Lamb, Timothy J., "Army Nonlethal Weapons/Scalable Effects Program: A Think Piece," *Military Police*, PB 19-03-1, April 2003, pp. 10–13.

———, "Emerging Nonlethal Weapons Technology and Strategic Policy Implications for 21st Century Warfare," *Military Police*, PB 19-03-1, April 2003, pp. 6–9.

Libicki, Martin C., David C. Gompert, David R. Frelinger, and Raymond Smith, *Byting Back—Regaining Information Superiority Against 21st-Century Insurgents: RAND Counterinsurgency Study—Volume 1*, Santa Monica, Calif.: RAND

Corporation, MG-595/1-OSD, 2008. As of October 13, 2008:
http://www.rand.org/pubs/monographs/MG595.1/

Mackinlay, John, and Alison Al-Baddawy, *Rethinking Counterinsurgency: RAND Counterinsurgency Study—Volume 5*, Santa Monica, Calif.: RAND Corporation, MG-595/5-OSD, 2008. As of October 13, 2008:
http://www.rand.org/pubs/monographs/MG595.5/

National Research Council of the National Academies, Committee for an Assessment of Non-Lethal Weapons Science and Technology, *An Assessment of Non-Lethal Weapons Science and Technology*, Washington D.C.: The National Academies Press, 2003.

New York City Police Department Police Academy, *City of New York Police Department Police Student's Guide: Use of Force*, New York, July 2005.

Nutley, Erik, L., *Non-Lethal Weapons: Setting Our Phasers on Stun? Potential Strategic Blessings and Curses of Non-Lethal Weapons on the Battlefield*, Air War College, Center for Strategy and Technology, Occasional Paper No. 34, August 2003.

Shupe, Paul K., "Nonlethal Force and Rules of Engagement," *Military Police*, PB 19-03-1, April 2003, pp. 43–48.

Smith, Rupert, *The Utility of Force: The Art of War in the Modern World*, New York: Knopf, 2007.

Underhill, Jeffrey L., *Are the Department of Defense Non-Lethal Weapon Capabilities Adequate for the 21st Century?* U.S. Army War College Strategy Research Project, Carlisle Barracks, Pa.: U.S. Army War College, 2006.

University of Florida Police Department, *Department Standards Directive: Use of Force*, March 2007.

U.S. Army and U.S. Marine Corps, FM 3-24S/MCWP 3-335, *U.S. Army/Marine Corps Counterinsurgency Field Manual*, Washington, D.C., December 16, 2006.

U.S. Army, U.S. Marine Corps, U.S. Navy, and U.S. Air Force, FM 3-22.40/ MCWP 3-15.8/NTTP 3-07.3.2/AFTTP(I) 3-2.45, *NLW Multi-Service Tactics, Techniques, and Procedures for the Tactical Employment of Nonlethal Weapons*, Washington, D.C., October 2007.

Wilkinson, Alec, "Non-Lethal Force," *The New Yorker*, June 2, 2008, pp. 26–33.